NEW YORK
CLASSICS

THE BOOK OF MY LIFE

GIROLAMO CARDANO (1501–1576) was born in Pavia, Italy. A professor of mathematics at Padua, and of medicine at Pavia and Bologna, he was the the author of more than a hundred books on subjects ranging from the natural sciences to medicine, history, and music.

ANTHONY GRAFTON teaches the history of Renaissance Europe at Princeton University. His books include *Joseph Scaliger*, *Cardano's Cosmos*, and *Bring Out Your Dead*.

HIERONYMVS CARDANVS MEDIOLANENSIS
Ætatis LXVIII

THE BOOK OF MY LIFE

(DE VITA PROPRIA LIBER)

GIROLAMO CARDANO

Translated from the Latin by

JEAN STONER

Introduction by

ANTHONY GRAFTON

NEW YORK REVIEW BOOKS

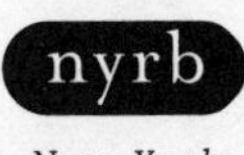

New York

To Mr. Waldo H. Dunn of the College of Wooster, to Mr. G.L. Hendrickson of Yale, to Miss Lucille Rand, my associate at the Dalton Schools in New York City, and especially to my sister, Miss Leah Stoner, also of the faculty at the Dalton Schools, I wish to acknowledge my indebtedness for the help and encouragement they gave me in preparing this translation.

—JEAN STONER
New York, October 1929

This is a New York Review Book
Published by The New York Review of Books
1755 Broadway, New York, NY 10019

Library of Congress Cataloging-in-Publication Data
Cardano, Girolamo, 1501–1576.
[De propria vita. English]
The book of my life = (De vita propria liber) / Girolamo Cardano ; translated from the Latin by Jean Stoner ; introduction by Anthony Grafton.
p. cm. — (New York Review Books classics)
ISBN 1-59017-016-4 (pbk. : alk. paper)
1. Cardano, Girolamo, 1501-1576. 2. Scientists--Italy--Biography. I. Title: De vita propria liber. II. Title. III. Series.
Q143.C3 A3 2002
509.2—dc21
2002003048

ISBN 1-59017-016-4

Book design by Lizzie Scott
Printed in the United States of America on acid-free paper.
10 9 8 7 6 5 4 3 2 1

September 2002
www.nybooks.com

CONTENTS

INTRODUCTION

GIROLAMO CARDANO dazzled readers across sixteenth-century Europe. His original and influential books dealt with medicine, astrology, natural philosophy, mathematics, and morals—to say nothing of devices for raising sunken ships and stopping chimneys from smoking. They won the attention of popes and inquisitors, Catholics and Protestants, theologians and playwrights. And nothing did more to enhance their appeal than the polished stories about Cardano's past that glittered like enticing gems in his most technical treatises, or the longer, dramatic retellings of his whole life story that he offered the public at suitable intervals.

Renaissance artists and writers specialized in self-scrutiny. The sixteenth century began with Dürer's hypnotically precise self-portrait, and approached its end with Montaigne's luminously introspective essays. But Cardano outdid all the rest. He pored over his past accomplishments and disasters with obsessive interest. He reread his many books so often, and with such delight, that eventually he explained the myth of Narcissus as an allegory of the scholar who loses himself in the pleasure inspired by his own writings. He also chronicled his own doings and undoings in works of every imaginable kind. Cardano composed a series of short bibliographical essays in imitation of the ancient medical writer Galen and the modern humanist Erasmus. In these, he described his books and their places in his life: numbers of pages, circumstances of composition, inspiring dreams.

Astrologers held that intervals of seven and nine years were especially significant. Cardano, accordingly, drew up, at intervals of about nine years, a series of versions of his own horoscope. In minutely specific, house-by-house commentaries, he told his readers about his stormy family relations, his digestion, his fluxes of urine, and his failures and successes as doctor and writer. He sprinkled many of his works on medicine and other subjects with vivid anecdotes of disease and cure, emergencies and autopsies among the rich and famous, drawn from his own practice as a doctor to the powerful in Milan, the Papal States, and Northern Europe. And in 1575, the last full year of his life, Cardano composed his most detailed autobiography of all, the richly textured, lurid, and sometimes eerie *Book of My Life*, which the French scholar Gabriel Naudé published, from a bad manuscript, in 1643, and which Jean Stoner translated into English in the 1930s. Stoner worked from a corrupt base text, and produced an imperfect version. But her translation reads well, gives a clear impression of both the author and his book, and has the great merit of existing.

Cardano's multiple self-portraits fascinated and alarmed the readers who scrutinized them, from the censors in the Holy Office to magicians in Germany and England. In this age of religious war and intellectual intolerance, courtly service providers like Cardano endured constant scrutiny, much of it hostile, from patrons and rivals alike. Safety lay in absolute reticence. Yet Cardano astonished—and horrified—readers by his frankness. He confessed in public that he had enjoyed the advice and visits of a familiar spirit—and that he had suffered years of sexual impotence despite his best efforts, that he lurched like an archetypal silly professor when he walked, and even that his servants took advantage of him. No wonder many readers—including Cardano's first editor, Naudé, and the great criminologist Cesare Lombroso—have been convinced that Cardano was mad, while others wondered if a devil had possessed him. *The Book of My Life* challenges, provokes, and amazes, even now.

Unlike most scholars, Cardano had many exciting experiences to record. Born in Pavia in 1501, he grew up in Milan and learned mathematics and astronomy from his father, Fazio Cardano. After studying in Padua and Pavia, he took a doctorate in medicine in 1525 and married in 1531. At first he had to struggle to make a living in Saccolongo, his wife's home, where he settled. As an illegitimate son, he could not gain entrance to the college of physicians in Milan, and made do by giving lectures on mathematics and practicing medicine on a modest, local level. In the 1530s, however, Cardano won the support of an important cleric, Filippo Archinto. Colleagues in Milan began to consult him on medical cases and astrological problems. He wrote a couple of pamphlets on astrology, one of which enabled him to bring off a real coup.

The distinguished Nuremberg publisher Johannes Petreius saw Cardano's pamphlet and sent the astronomer Georg Joachim Rheticus, who worked as one of his agents, to seek the author out. Eventually Petreius published a long series of Cardano's books—most notably his pioneering study of algebra, *The Great Art*, the first treatise in academic Latin to survey the new Italian mathematics of the sixteenth century. This strikingly innovative work, suitably adorned with a portrait of the author and a phosphorescently enthusiastic blurb, appeared on Petreius's list in 1545, only two years after Copernicus's *The Revolutions of the Heavenly Spheres*. In its own way, *The Great Art* had equally radical effects, revolutionizing the teaching of mathematics and involving Cardano himself in sharp debates about intellectual priority and plagiarism, since Niccolò Tartaglia pointed out that Cardano had violated a promise not to publish Tartaglia's methods. The illegitimate Milanese physician had become a European celebrity. Thanks to a professorship at Pavia, he prospered on the local level as well.

Like other polymaths in that age of universal savants, Cardano wrote, on a dazzling variety of fields, more books than any modern could hope to read. His treatises and commentaries on medicine, natural philosophy and magic, and astrology reached an immense audience, not only in Catholic Europe but also in the heartlands of Protestantism, like Wittenberg itself. Martin Luther made fun of the horoscope Cardano drew up to explain why he had rebelled against the Catholic Church. But Luther's closest friend, Philipp Melanchthon, read Cardano's work with care and taught it to his students. The arc of Cardano's career curved upward even more steeply in the 1550s. He followed his books northward in the mid-1550s, in order to treat Archbishop Hamilton of St. Andrews, in Scotland, for a cardiopulmonary illness. The archbishop's personal physician, a reader of Cardano, had noticed the author's claim that he could cure such diseases. In the meantime, Cardano had lost faith in his original method. Still, he accepted the invitation, and the archbishop survived—for fifteen years, to be executed when Scotland became a Protestant country.

In the course of his travels, Cardano realized that he had somehow metamorphosed into a grandee. In both France and England, the good and the great asked to meet him, had him draw up their horoscopes, and introduced him to other wielders of power and influence. When Cardano arrived in London, he interviewed the scholar John Cheke, the young King Edward, and other grandees, in order to produce accurate horoscopes based on data they themselves provided. His skills brought him into high company: the English astrologer and mathematician John Dee inspected a magic gem with Cardano in the house of the French ambassador to London, and the two of them also investigated a perpetual motion device.

Cardano's in-your-face frankness and penchant for risky claims led him into more than one disaster. In 1554, he published his horoscope for Edward of England, which predicted that its subject would marry, rule his country for some time, and live a reasonable span. It appeared in print just after the boy king died, forcing Cardano to make lame and lengthy efforts at self-justification. Three years later, another of his works provoked the longest and most vitriolic book review in the annals of literature—the nine-hundred-page diatribe that an Aristotelian natural philosopher and medical man, Julius Caesar Scaliger, directed against Cardano's treatise *On Subtlety*. Scaliger found a tempting target in every one of Cardano's characteristic boasts (indeed, in every one of his flowery metaphors), and picked them off with contemptuous ease. His massive, compulsively readable polemic—which he described as only one of fifteen that he planned to publish—became an assigned text in many universities. Cardano's book continued to sell, and he wrote a sharp reply—but he clearly came off the loser.

Literary catastrophes at least confirmed that Cardano was worth attacking. But his personal life changed decisively for the worse in the 1550s. His wife died, and he continued to feel lonely and isolated, even as he built up the sort of elaborate household that a man of substance deserved. Nothing came easy to Cardano. Illegitimacy and poverty slowed his rise. Tormented as a young man by sexual impotence, in middle age he found the dangers of medical practice even more terrifying. In a time of plague and high mortality, doctors were always vulnerable. Cardano had more rivals than most, and they did their best to discredit him for every patient's death.

His troubles began to arrive in battalion strength, moreover, just when his career reached its peak. He moved to the ancient university of Bologna in 1562, hoping to find a haven from his colleagues' plots and attacks. But in Bologna too, his colleagues criticized him in public and undermined him in private. His enemies advised his students to study with others, assigned his favorite lecture times to rival teachers, and then complained about his courses' low enrollments. His eldest son, whom he adored and saw as endowed with immense promise, was arrested, condemned, and executed for poisoning his wife, and Cardano found himself forced to expel his younger son from his house as a thief.

At the same time, the larger social and cultural climate turned increasingly bleak. During his student years in the 1510s and 1520s, Cardano had seen Italy lose its political independence, as French and Imperial armies fought their way up and down the peninsula. But these decades had also been a time of hope and renewal. Bearded prophets stalked city streets and piazzas, calling for a new order in the Church. Peddlers distributed the pamphlets of Savonarola and his followers, not to mention those of Luther and Melanchthon. Ignatius Loyola and a host of others—including Cardano's patrons—created new forms of religious life within the Catholic Church, founding new religious orders and reconfiguring Catholic dioceses. Cardano shared the hopes of his generation. He praised Erasmus and other humanists, some of them Protestants, who tried to restore the original message of Christianity. In 1534, he even published an astrological pamphlet that predicted that the church would soon undergo a radical transformation.

As Cardano aged, however, the winds blew colder. The new doctrinal rigor called for between 1545 and 1563 by the Council of Trent—and the new censorship imposed by the Congregation of the Holy Office, through the Index of Forbidden Books—began to hamper free investigation and expression of ideas. Cardano, with his talk of angels and daemons, his provocative books and challenging language, naturally seemed suspect. In 1570 he was arrested and tried and underwent a short period of imprisonment. The Bolognese Inquisition forbade him to teach or publish, but released him fairly soon, and in March 1571 he made his way to Rome. Still passionately articulate, he impressed the Roman physicians and managed to win a papal pension when Gregory XIII succeeded Pius V.

Cardano remained an imposing figure—but he was a deeply frustrated man. True, he enjoyed revising his work (one of his cleverest inventions offered writers an easy way to transform their texts without copying them over). But he hated it when the censors forced him to draft corrections and additions to his early works, and he had little success at convincing the authorities that he was innocent or penitent. In another late work, he remarked that impiety and insanity were both serious charges, but impiety was "more dangerous, especially in these times." Cardano the astrologer sat in his bare rooms in Rome, decorated only with a banner claiming "Time is my Possession," and knew that in sad fact his times were out of joint. Still famous and respected when he died in 1576, he might have ended on a pyre like Giordano Bruno if he had lived somewhat longer.

Cardano spent his last year, in large part, writing *The Book of My Life*. It seems a strange book now, and already seemed one when it first appeared. Fluid, chaotic, endlessly digressive, it veers from one subject to another, from organizational scheme to organizational scheme. At times, Cardano narrates his life more or less in order. At times, he lists his publications or his friends. Most often, he analyzes one aspect after another of his life, character, or experience, examining his digestion and his dreams, his marriage and his medical practice, as if he were composing in retrospect the sort of elaborate, luxurious horoscope that he had drawn up for his wealthy clients, which predicted their bodily and spiritual health, domestic relations, and careers, topic by topic.

The Book of My Life, which begins with the astrological details of Cardano's birth, actually grew from the full commentaries he had written on his own horoscope in previous decades. Though the book eventually became far longer than any of Cardano's other horoscopes, it adopted their mosaic-like form—a form that other biographers and autobiographers of the time also found suggestive and rewarding. John Aubrey's charming *Brief Lives*, for example, grew from his own collection of horoscopes—and reveal a similar notion of how lives are shaped by temperaments and bodies, and those in turn by the cosmos.

From the start, Cardano tried to meld this technical form of analysis with a much more traditional form of literary self-presentation. As a successful writer who had worked hard to orchestrate the success of his books, and who considered at least some of his works as divinely inspired, he joined the ranks of poets and philosophers, from Petrarch onward, who described their own lives as part of an effort to preserve their memory from the gnawing tooth of time. Petrarch emulated Roman poets like Ovid when he addressed himself, in a formal letter, "to Posterity." He explained that readers centuries later might feel as much curiosity about him as he did about the Latin poets. Many other learned Latinists in their turn wrote autobiographies or dictated extensive autobiographical details to friends. Cardano was hardly the only Renaissance writer to portray himself, at times, as a heroic figure, or to give readers striking details about his financial disasters and emotional depressions. To that extent his book began, at least, as a normal literary enterprise.

In this case as elsewhere, however, Cardano found that too much could never be enough. He went far beyond the bounds of normal discretion when he revealed his experiences with a supernatural being, and far beyond the bounds of normal introspection when he scrutinized every conceivable omen, prodigy, or astrological sign that could have had some oblique connection with his career. As a medical man and an astrologer, Cardano deeply believed that a web of connections and sympathies bound each individual—his bodily organs, his temperament, even his tastes and interests—to the stars that gave life to everything on earth and to the supernatural beings that populated the heavens. Any sign, however small—the sound of buzzing with no bees to account for it, the smell of wax in the absence of burning candles—could give the astute reader of the world the clue he needed to establish a connection or make a prediction, the indication he could use to change his diet or his habits and avoid a terrible disease.

Reading Cardano's autobiography enables one to feel, intimately, what it was like to inhabit a world designed by a divine intelligence down to the smallest details and strewn with clues to what these meant—a world that hummed with hidden but vital messages, that the scholar spent his life deciphering. Like one of the Italian gardens of the time, with their deep grottos, monstrous sculptures, and hidden water traps that drenched those unwary enough to sit on an inviting bench, Cardano's book was designed to surprise and delight—and dismay—its readers, to make them feel the wonder with which experience continually inspired him.

If Cardano saw himself as the largely passive prey of the cosmos, he never saw autobiography as a passive act of recording. Like many of his contemporaries, he constantly hoped to improve and discipline himself. Renaissance readers knew that ancient philosophers had offered not only systems of ideas, but rules for living, and they tried to internalize these. They read the Stoics, for example, to learn how to survive the worst blows of fate. Marcus Aurelius—whose own autobiographical work became available in Latin not long before Cardano wrote—gave a splendid example of how the philosopher could use introspective writing to examine, and correct, his own character. At the same time, a host of new self-help books instructed Cardano and his contemporaries on what to eat, how much exercise to take, how to dance, and even how to walk. Cardano constantly tried, as he recorded his own experiences, to work out which of these systems had helped him, and which had not. With his customary air of frankness, he told readers how he had broken the rules of prudence, good taste, and emotional restraint—and yet scored many successes in his career and survived calamities and ridicule in his private life.

The Book of My Life, with its intimate record of despair and exaltation, crisis and triumph, confrontation and debate, recounts the complex history of a tortured soul, one that constantly tried to shape the body it inhabited and the desires that ravaged it. If Cardano never entirely succeeded in attaining the cold humanist self-control that he wished for, he also never stopped functioning, even when his son died or the authorities persecuted him. To that extent, Cardano's book is more than self-advertisement or applied astrology. It is his scrupulous, heroic, and necessarily incomplete effort to instruct the world by turning his soul into a case history. Cardano's unsparing courage, his willingness to examine his own depths, and his passion to find the meaning in the apparently insignificant details and transactions of daily life, remind one of Freud. But Cardano, unlike Freud, made no effort to conceal his own identity or those of his friends and clients: his theories and explanations remained tightly bound to the vivid, particular stories he told so well. These made *The Book of My Life* compelling reading for Enlightenment philosophes and Biedermeier burghers, and they continue to charge it with interest now.

—ANTHONY GRAFTON

PROLOGUE

THIS Book of My Life I am undertaking to write after the example of Antoninus the Philosopher,[1] acclaimed the wisest and best of men, knowing well that no accomplishment of mortal man is perfect, much less safe from calumny; yet aware that none, of all ends which man may attain, seems more pleasing, none more worthy than recognition of the truth.

No word, I am ready to affirm, has been added to give savor of vainglory, or for the sake of mere embellishment; rather, as far as possible, mere experiences were collected, events of which my pupils, especially Ercole Visconte, Paulo Eufomia, and Rodolpho Selvatico[2] had some knowledge, or in which they took part. These brief cross-sections of my history were in turn written down by me in narrative form to become this, my book.

Gaspare Cardano, a physician by profession, and one of my kinsmen and pupils, had attempted this very thing a few years ago, but, being overtaken by death, was unable to complete his work. Also a Jew,[3] a man without the justification of having been so much as a magistrate, felt at liberty to carry out the same idea, and did so without criticism. Accordingly, even if nothing of great moment has befallen me, I feel that I too have certainly had many noteworthy experiences. Nor has it escaped me that Galen[4] endeavored to give us an account of his life. Owing to the fact that to himself it seemed more appropriate to reveal one incident and another here and there throughout his works, it has happened that because of the indifference of students, no classic writer has even attempted to put it in order. My autobiography, however, is without any artifice; nor is it intended to instruct anyone; but, being merely a story, recounts my life, not tumultuous events. Like the lives of Sulla, of Caius Cæsar, and even of Augustus, who, there is no doubt, wrote accounts of their careers and deeds, urged by the examples of the ancients, so, in a manner by no means new or originating with myself, do I set forth my account.

THE BOOK OF MY LIFE

1.

NATIVE LAND AND FOREBEARS

MY NATIVE country is the Duchy of Milan; the town in which the family of the Cardani had its origins is twenty-four miles distant from the city. From Gallarate it is only seven miles.

My father Fazio was a jurisconsult. My grandfather was Antonio, my great-grandfather another Fazio, my great-great-grandfather, Aldo. The first Fazio, my great-grandfather, had three sons: Giovanni, Aldo II, and Antonio, who became my grandfather. The sons of Antonio were Gottardo, Paolo, a jurisconsult and prelate, Fazio, my father, and an illegitimate son named Paolo. Now there are alive, from the same family, almost thirty kinsmen. Whether, however, the family of the Cardani is a line in itself, or whether, as certain people think, it is a branch of the Castiglioni, surely it has been both noble and ancient. From the year 1189 a Milone Cardano was prefect of Milan for seven years and eight months by the ecclesiastical as well as the civil calendar. And he presided not only at civil suits, but even, like other Chiefs of State, at criminal trials; he had charge, also, of cases in other cities which were under Milanese jurisdiction; and cases throughout the whole region. Among these cities was Como. Milone had secured this high post of honor through the Archbishop Crivelli when the latter had been hurriedly advanced to the Pontificate as Pope Urban III.[1]

There are those who would like to have it that Francesco Cardano, leader of the military troop of Matteo Visconti, belongs to our branch of the family. If we are a branch of the

Castiglioni, however, we are still more illustrious because of Pope Celestino IV[2] who sprang from this same stock.

All our forebears have, moreover, been longevous. The sons of the first Fazio lived ninety-four, eighty-eight, and eighty-six years. The two sons of Giovanni, my great-uncle, were Antonio, who lived eighty-eight years, and Angiolo, who lived ninety-six years. As a boy I saw this decrepit old man. Giacomo, an only son of the second Aldo, passed seventy-two years. Gottardo, my paternal uncle, lived eighty-four years, and him also I saw. My father lived eighty years. That cousin Angiolo, the one who reached his ninety-sixth year, begot sons —infants feeble as if with their father's senility—even when he himself was in his eightieth year, and after his eightieth year he recovered his sight. The eldest of these sons has lived to be seventy, and I hear that some of his children became giants; indeed, I saw some of them long ago.

There were, besides, from the family of the Micheri (my mother Chiara's family), my grandfather Giacomo, who lived seventy-five years, and his brother Angiolo, who was in his eighty-fifth year when I was a little boy. He himself told me so.

Learning and a singular integrity of life were the common characteristics of my father, my paternal uncle, and my mother's father, while the last named and my own father were men gifted with great old age and exceptional skill in mathematics. This same grandfather was in prison at a slightly different age from my own when I was imprisoned, although each of us was in his seventieth year.

Five other branches of the family of the Cardani existed, coming from our Aldian stock: that of Antonio, dating from 1388; the Gasparini, from 1409; that of Rainieri, 1391; and Enrico's, the oldest, dating from 1300, whose descendants, Berto and Giovanni Fazietto, are about of equal age. There was also Guglielmo—it is not certain just when he flourished—and his sons, Zolo, Martino, and Giovanni, who lived at Gallarate.

2.

MY NATIVITY

ALTHOUGH various abortive medicines—as I have heard—were tried in vain, I was normally born on the 24th day of September in the year 1500, when the first hour of the night was more than half run, but less than two-thirds.[1]

The most significant positions of the horoscope were as I have indicated in the eighth nativity of the supplement to the four sections of my *Commentaries on Ptolemy*.[2] I have taken into consideration, therefore, that both luminaries were falling in the angles, and neither was applying to the ascendant, inasmuch as they were posited in the sixth and twelfth houses. They might have been in the eighth house, subject to the same condition, for the latter house descends and is not an angular house; a planet therein could rather be said to be falling from the angle. And although the malefics[3] were not within the angles, nevertheless Mars was casting an evil influence on each luminary because of the incompatibility of their positions, and its aspect was square to the moon.

Therefore I could easily have been a monster, except for the fact that the place of the preceding conjunction had been 29° in Virgo, over which Mercury is the ruler.[4] And neither this planet nor the position of the moon or of the ascendant is the same, nor does it apply to the second decanate of Virgo; consequently I ought to have been a monster, and indeed was so near it that I came forth literally torn from my mother's womb.

So I was born, or rather taken by violent means from my mother; I was almost dead. My hair was black and curly. I was revived in a bath of warm wine, which might have been fatal to any other child. My mother had been in labor for three entire days, and yet I survived.

Besides, to return to the horoscope, since the sun, both malefics, and Venus and Mercury were in the human signs, I did not deviate from the human form. Since Jupiter was in the ascendant and Venus ruled the horoscope, I was not maimed, save in the genitals, so that from my twenty-first to my thirty-first year I was unable to lie with women, and many a time I lamented my fate, envying every other man his own good fortune. Although Venus was, as I have said, ruler of the whole nativity, and Jupiter in the ascendant, unfortunate indeed was my destiny; I was endowed with a stuttering tongue and a disposition midway between the cold and the harpocratic—using Ptolemy's classification—that is to say, gifted with a kind of intense and instinctive desire to prophesy. In this sort of thing—it is called prescience, to use a better expression—as well as in other methods of divining the future, I have been clearly successful at times.

Because Venus and Mercury were beneath the rays of the sun, and to this luminary owing their combined strength, I could, accordingly, have escaped some of the consequences of entering my existence with a genesis—as Ptolemy calls it—so wretched and luckless, had the sun itself not been directly in its fall, cadent in the sixth house and removed from its own exaltation.[5] I was gifted, therefore, with a certain cunning only, and a mind by no means at liberty; my every judgment is, in truth, either too harsh or too forbidding.

In a word, I shall say that I am a man bereft of bodily strength, with few friends, small means, many enemies—a very large part of whom I recognize neither by name nor by face—a man without ordinary human wisdom, inclined to be faulty of memory, though rather better in the matter of foreseeing events. Yet withal, my condition in life, to be deemed lowly if one considers my family and my betters, is, for some unknown reason, regarded as honorable and worthy of emulation among these same.

On my birthday Augustus was born long ago[6]; throughout the whole Roman Empire a new regime began. On my birthday also, Ferdinand, the most gracious King of Spain, and Isabel, his wife, first sent out to sea the fleet which discovered the whole occidental world.[7]

3.

CERTAIN TRAITS OF MY PARENTS

MY FATHER went dressed in a purple cloak, a garment which was unusual in our community; he was never without a small black skullcap. When he talked he was wont to stammer. He was a man devoted to various pursuits. His complexion was ruddy, and he had whitish eyes with which he could see at night; up to the very last day of his life he had no need to use glasses. This admonition was forever upon his lips: "Let every spirit praise the Lord, who is the source of all virtue." Because of a wound received on the head in his youth, some pieces of bone had been removed, and he consequently could not remain long without his cap. From his fifty-fifth year on he lacked all his teeth. He was well acquainted with the works of Euclid; indeed, his shoulders were rounded from much study. My eldest son was much like my father in features, eyes, shoulders, and manner of walking; but the lad was quicker with his tongue, perhaps on account of his age.

The only very intimate friend of my father's, in spite of a decidedly different calling, was Galeazzo Rosso, this last being the family name. He died before my father. There was also the Senator Gianangelo Selvatico, a student and ward of my father's.

The former was a smith who became acquainted with my father through a certain community of habits and similar intellectual pursuits. He it was who had discovered the screw of Archimedes before the works of Archimedes had been published. He invented a method of so tempering swords that they could be bent like lead, and yet would almost cut iron as if it were wood. What is more, he wrought steel breastplates which resisted the shots of the firearms which the regimental soldiers carried; one plate alone sufficed to withstand a blow of five times ordinary force, and suffered scarcely a dent. I myself was frequently an onlooker at this experiment when a young man.

My mother[1] was easily provoked; she was quick of memory and wit, and a fat, devout little woman. To be hasty-tempered was a trait common to both parents; they were not consistent, even in love toward their child. Yet they were indulgent, to the extent that my father permitted me, indeed bade me, not to arise from my bed until after eight o'clock, a practice which greatly benefited my life and health. My father seemed better—if I dare say as much—and more loving than my mother.

4.

A BRIEF NARRATIVE OF MY LIFE FROM THE BEGINNING TO THE PRESENT DAY—THAT IS, THE END OF OCTOBER, 1575

IF SUETONIUS by any chance could have noticed the method of this chapter, he might have added something to the advantage of his readers[1]; for there is nothing, as the wise men say, which may not in some manner be unified.

Having been born at Pavia, I lost, in the very first month of my life, my wet-nurse on the day when she fell ill, so they tell me, of the plague[2]; and my mother returned to me. Five carbuncles came out on my face in the form of a cross, and there was one on the end of my nose; in these same spots as many swellings, commonly called variolas, appeared again after three years.

When my second month was not yet run, Isidoro dei Resti, of Pavia, took me naked from a warm vinegar bath and gave me over to a wet-nurse. The latter carried me to Moirago, a village seven miles away from Milan in direct line from our town of Pavia, through the town of Binasco. There my body wasted, while the belly grew hard and swollen. When the reason was known—that my nurse was pregnant—I was transferred to a better nurse, who suckled me until I was three.

In my fourth year I was taken to Milan. Here, although I received kindly treatment (except on occasions when I was unjustly whipped by both my father and my mother) at my mother's hands and from my mother's sister, my Aunt Margherita—a woman, I think, without any real animosity—I was many times ill to the very verge of death. At length when I was just seven years old—my father and my mother were not living together then—and really justly deserved an occasional whipping, they decided to refrain from this punishment.

My bad luck had by no means deserted me, but had simply changed my misfortunes without removing them. Having rented a house, my father took me, my mother, and my aunt home with him; and, thereupon I was ordered to accompany him daily in spite of my tender age and frail little body; I was pushed from a very tranquil existence into a round of tiring and constant exertions. Consequently at the beginning of my eighth year I fell ill of a dysentery and fever. An epidemic of this disease, if not a very plague of it, was then abroad in our city; and I had, besides, eaten a quantity of green grapes. Barnabo Croce and Angelo Gira were summoned. Before I was pronounced out of danger both my parents and my aunt had wept over me as if I were actually dead. Thereupon my father, who was a man devout at heart, wished rather to seek the divine help of Saint Jerome, with a vow commending my welfare to him, than to call upon the demon which he openly confessed attended him as a familiar spirit,[3] a relationship the significance of which I steadfastly refrained from investigating. I was convalescing when the French, after the Venetians had been defeated in the region of the Adda, were celebrating a triumph,[4] which I was allowed to view from my window. On my recovery the duty of accompanying my father and the constant tasks were given up.

But the wrath of Juno was not yet satiated. I was not entirely well from my sickness when I fell down the stairs with a hammer while we were living in the Via dei Maini. The hammer struck the upper part of my forehead on the left side, and I was wounded with somewhat serious results to the skull; the scar I have to this very day, and shall have always. I was just getting over this, and was sitting in the doorway of my home, when a stone, as wide and as long as a nut, but thin like a slice of bark, came hurtling down from the top of a high house adjoining our own, wounding me again on the left side where my hair was quite thick.

When I was about ten years old, my father moved from that apparently accursed house, and took another away from the immediate neighborhood but in the same street. There I remained three years. My luck had not yet changed, for again my father, stubbornly insistent, had me accompany him as his page. I shall not say that he acted with severity; this seemed to have been done rather because it was divinely intended, than because of any fault of my father, as the reader may believe from events which ensued. Besides, my mother and my aunt heartily agreed that he should use me in this way. However, he dealt with me far more leniently than formerly; for during this period, he had with him two nephews, one after the other, because of whose services mine became so much lighter that it was not always necessary to accompany him, and the work was not so irksome when I did. Thereafter I was always his page. In the meantime we again moved, and at length, toward the end of my sixteenth year, we took up residence near the bake-shop of Bossi in the house of Alessandro Cardano.

My father had two nephews, sons of his sister: one, Evangelista, belonging to the order of the Franciscans, and almost seventy years old; the other Ottone Cantoni, a collector of revenue and a rich man. The latter, before dying, had wished me to be the heir of his entire estate, but my father forbade it, saying the money was ill-gotten; and so the property was distributed at the discretion of the brother who survived him.

At the end of my nineteenth year I went to the academy at Pavia,[5] in the company of Gian Ambrogio Targa; there I remained the following year also, but without a companion. When I was twenty-one I returned to Pavia with the same friend. In this year I debated in public, and gave instruction in Euclid at the Gymnasium. A few days after this I lectured on dialectics, and also in a course in elementary philosophy: first for Fra Romolo Servita, and later in several classes for a certain physician named Pandolfo.

During my twenty-second year I remained at home on account of the fighting which brought everything else to a standstill in our district. But shortly after the beginning of 1524 I went to Padua.[6] Luckily, toward the end of the year, that is, in the month of August, I went to Milan with Giangelo Corio, where I found my father in the clutches of an acute sickness. He, anxious for my welfare rather than his own, ordered me to return to Padua, because he knew that I had received the Venetian degree in Arts,[7] the Baccalaureate, as they say. When I had returned, I received letters saying that my father had died, after he had abstained from food for nine successive days. He had begun to fast on August 20th, a Saturday, and died on the 28th.

Toward the end of my twenty-fourth year I was made Rector of the College.[8] At the end of the following year, I was granted the degree of Doctor of Medicine. The first honor I secured by a majority of one, the ballots having twice been cast. As a candidate for the second I lost twice, since forty-seven votes were cast against me. The third trial was the last; I came out first place, since nine ballots were cast against me, exactly as many as had been for me in the previous elections, leaving forty-seven votes in my favor.

I am not blind to the insignificance of these facts. I have simply put them down in the order in which they happened, so that they may give satisfaction to me as I read them. I do not record such personal items for others' eyes. At the same time, if there chance to be some who will even deign to read, may these note that the beginning and outcome of important events are not always evident; whether because they are such as commonly happen to others rather than ourselves, or to us when we are not aware of their occurrence.

After my father's death, when my term of office had been completed, I settled, early in my twenty-sixth year, in the town of Sacco, a village ten miles from Padua and twenty-five from Venice. This I did, acting upon the advice and suggestion of Francisco Buonafede, a physician of Padua. He was friendly to me out of the very abundance of his exceptional high character; for he had nothing to gain from his kindness to me, who had not even been attending his public lectures.

I remained in Sacco while my country was being devastated by all manner of evils. A terrible plague visited it in 1524, and twice the governorship changed hands. In 1526 and in 1527 the crops were almost fatally scanty, so that scarcely with money could a signed requisition for grain be redeemed. The taxes were insupportable. In 1528 we were again besieged by disease and pestilence, which were fairly innoxious, as the country was already ravaged far and wide.

I returned in 1529 to my native city, after the tumult of war had somewhat abated. There I was rejected by the College of Physicians[9]; nor was I able to make any headway in my lawsuit with the Counts Barbiani. But, as my mother was bitterly depressed, I returned to my own little town of Sacco, scarcely in such good health as I had left it. For, what with sickness, hard work, anxieties, and presently, added to these, a cough, loss of flesh, and empyema accompanied by loathsome discharges, I reached the point whence one scarcely dares hope to return to sound health. At length I was freed from my disorder under a vow to the Blessed Virgin, and toward the end of my thirty-first year I married Lucia Bandarini of Sacco.

Four things, moreover, have I observed occurring to this very day: that enterprises I have undertaken before full moon have turned out successfully, although my act was not always premeditated; again, about the time when other men are wont to abandon hope of happiness I have entered into it; furthermore, as I have elsewhere said, Fortune, often on the very verge of desertion, has stood by me; finally, up to my sixtieth year nearly all my journeys were undertaken in the month of February.

My wife, after a second miscarriage, bore me two sons, and, between them, a daughter.

In the following year, toward the end of April, I moved to Gallarate where I lived nineteen months, during which time I was restored to good health. And I ceased to be poor because I had nothing left. At Milan, finally, through the kindness of the prefects of the great Xenodochium,[10] and also by the aid of Filippo Archinto, *vir illustris*, and at that time an orator of merit, I began to lecture publicly in mathematics. That was when I was about thirty-four years old. Two years later I was invited to teach medicine at Pavia, an offer which I did not accept, for there seemed little hope of receiving pay for the work.

In the year 1536 I went to Piacenza, summoned to the Pope by letters from the Prelate Archinto,[11] who had not yet been made priest; my journey bore no fruit. In addition, the Viceroy of the French was urgently soliciting favor in my behalf at the instance, as I later heard, of the nobleman Louis Birague, Commander of the infantry of the French King in Italy. Since Viceroy Brissac was a man singular for his devotion to gallantry and to letters, he offered many inducements; but nothing came of it.[12]

The following year, 1537, I again made application to the College of Physicians in Milan, and again was flatly rejected. In 1539, however, I was, contrary to every expectation, received; many had withdrawn their objections, and two excellent friends, Sfrondrato and Francesco della Croce,[13] had come to my aid.

Accordingly, from the year 1543 I lectured on Medicine in Milan; and the year after—the year in which my house collapsed—I lectured in Pavia, although the lectures were practically unattended. Since the stipends were not forthcoming, I discontinued these lectures in 1544. I remained at Milan with my eldest son who was at that time completing his eleventh year; my daughter was in her ninth year, and Aldo was a little two-year-old.

About that time, Cardinal Morone[14]—whom I name for honor's sake—in the summer of 1546, made offers that could not reasonably be spurned. But I, who, as I have said, am gifted with foresight, said to myself, "The high pontiff[15] is in his dotage; he is practically a crumbling wall. Why should I leave certainties for uncertainties?" To be sure, I was not cognizant at that time of the high character of Morone, nor of the glories of the Farnesi. I had already from 1542 become involved in friendship with the Prince d'Iston,[16] who acted as my patron, and who would have given much more than I was willing to receive. But after this summer, 1546, I returned to my task of lecturing.

The next year through the instance of Andreas Vesalius,[17] a man of highest standing, and my friend, an offer of 800 gold crowns yearly was made to me by the King of Denmark. This I was not eager to receive—although he even offered living expenses—not only on account of the severity of the climate in that region, but also because the Danes are given to another way of worship. I felt that perhaps I might not be very welcome there, or else might be forced to abandon the doctrines of my country and my forebears.

In my fiftieth year, because the stipend from the University of Pavia was not paid, I remained at Milan. During the month of February 1552, an occasion for going to Scotland was offered me.[18] I received before departing 500 gold crowns of France; and upon my return 1,200. I was gone 311 days; and had I wished to remain in Scotland, I would have received a much larger sum.

From January 1st, 1553, to October 1st, 1559, I continued to be engaged at Milan. During this period I rejected some even more generous inducements, one of these being from the French king. But I feared to offend the house of the Cæsars, because the latter were at war with the French.[19] Shortly after my return to Milan, a second offer came—this from the Duke of Mantua, his uncle, Don Ferrando, being the intermediary. The Queen of Scotland[20] also sought me with promises munificent, but too remote for certainty. I had treated her half-brother, the state of whose health stood in some hope of betterment. After I had effected a cure upon him she had been influenced by my practice, and by esteem.

Shortly after my return to Pavia in the year 1559, the occasion of my son's death came about.[21] Thereafter I endured my very existence, until 1562, when, summoned to Bologna,[22] I went there and was in uninterrupted employment until the year 1570. On the 6th of October of that year I was imprisoned.[23] There, beyond the fact that I had lost my liberty, I was civilly treated. On December 22, 1570, at the same hour and on the same day in which I had been imprisoned—I think it was a Friday—about evenfall I returned to my home. But this place was as yet a prison to me. The first term of incarceration was seventy-seven* days, which, together with the eighty-six days in my own home, made in all 162 days.

I stayed on in Bologna during 1571 until the end of my seventieth year, at the close of the month of September. Thence I came to Rome on the day of the celebrated victory against the Turks, October 7, 1571.[24]

At the present, to be exact, four years have passed since I entered the city, and five since my incarceration. I have passed my days as a private citizen, except for my reception by the College of Rome, on the 13th of September. The Pope is my patron in the matter of pension.[25]

---

*An error in computation; there are seventy-six days between dates.

5.
STATURE AND APPEARANCE

I AM A man of medium height; my feet are short, wide near the toes, and rather too high at the heels, so that I can scarcely find well-fitting shoes; it is usually necessary to have them made to order. My chest is somewhat narrow and my arms slender. The thickly fashioned right hand has dangling fingers, so that chiromantists have declared me a rustic; it embarrasses them to know the truth. The line of life upon my palm is short, while the line called Saturn's is extended and deep. My left hand, on the contrary, is truly beautiful with long, tapering, well-formed fingers and shining nails.

A neck a little long and inclined to be thin, cleft chin, full pendulous lower lip, and eyes that are very small and apparently half-closed; unless I am gazing at something... such are my features.[1] Over the eyebrow of my left eye is a blotch or wart, like a small lentil, which can scarcely be noticed. The wide forehead is bald at the sides where it joins the temples. My hair and beard were blonde; I am wont to go rather closely clipped. The beard, like my chin, is divided, and the part of it beneath my chin always was thick and long, seeming to have a more abundant growth thereunder. Old age has wrought changes in this beard of mine, but not much in my hair.

A rather too shrill voice draws upon me the censure of those who pretend to be my friends, for my tone is harsh and high; yet when I am lecturing it cannot be heard at any distance. I am not inclined to speak in the least suavely, and I speak too often.

I have a fixed gaze as if in meditation. My complexion varies, turning from white to red. An oval face, not too well filled out, the head shaped off narrowly behind and delicately rounded, complete a picture so truly commonplace that several painters who have come from afar to make my portrait[1] have found no feature by which they could so characterize me, that I might be distinguished. Upon the lower part of my throat is a swelling like a hard ball, not at all conspicuous, and coming to me as an inheritance from my mother.

6.
CONCERNING MY HEALTH

MY BODILY state was infirm in many respects: by nature; as the result of several cases of disease; and in the symptoms of weakness which displayed themselves.

My head is afflicted with congenital discharges coming at times from the stomach, at times from the chest, to such an extent that even when I consider myself in the best of health, I suffer with a cough and hoarseness. When this discharge is from the stomach, it is apt to bring on a dysentery and a distaste for food. More than once I believed I had had a touch of poison, but I shortly and unexpectedly recovered.

Another trouble was a catarrh or rheum of the teeth, through the effects of which I began to lose my teeth, several at a time, from the year 1563 on. Before that I had lost but one or two. Now I have fourteen good teeth and one which is rather weak; but it will last a long time, I think, for it still does its share.

Indigestion, moreover, and a stomach not any too strong were my lot. From my seventy-second year, whenever I had eaten something more than usual, or had drunk too much, or had eaten between meals, or eaten anything not especially wholesome, I began to feel ill. I have set forth a remedy for the foregoing in the second book of my treatise *On Guarding the Health.*

In my youth I was troubled with congenital palpitation of the heart, of which I was absolutely cured by medical skill. I had hemorrhoids, also, and the gout, from which I was so nearly freed that I was more frequently in the habit of trying to call it back when it was not present, than of getting rid of it when I had it.

I ignored a rupture, another weakness, in its early stages; but later, from my sixty-second year on, I greatly regretted that I had not taken care of it, especially since I knew it to be an inheritance from my father. In the case of this rupture, something worthy of note occurred: the hernia had started from either side, and although it was neglected on the left side, was eventually healed completely in that part by natural processes. The right side, more carefully treated with ligatures and other attentions, grew worse. A cutaneous itching annoyed me constantly, now in this part, now in that.

In 1536 I was overtaken with—it scarcely seems credible—an extraordinary discharge of urine; and although for nearly forty years I have been afflicted with this same trouble, giving from sixty to one hundred ounces in a single day, I live well. Neither do I lose weight—that I wear the same rings is evidence of this; nor do I thirst inordinately. Many others, seized that same year by a similar disease, and who did not seek a remedy, held out much longer than those who sought medical aid.

The tenth of these infirmities is an annual period of sleeplessness lasting about eight days. These spells come in the spring, in the summer, in the autumn, and in the winter; so that almost a whole month, rarely less, is spent yearly, and sometimes two. This I am wont to cure by abstaining from certain kinds of food, especially heavy food, but I do not diminish the quantity. This insomnia has never missed a year.

Several actual cases of sickness overtook me during my life. In the second month of my life I had the plague. The next serious illness occurred in or about my eighteenth year[1]; I do not recall the exact date, other than that it happened in the month of August. I went almost three days without food, and spent the time wandering about the outskirts of the town, and through the gardens. When I returned home at nightfall, I pretended that I had dined at the home of my father's friend Agostino Lanizario. How much water I drank in those three whole days I cannot truthfully say. On the last day, because I was not able to sleep, my heart palpitated wildly and the fever raged. I seemed to be on the bed of Asclepiades[2] in which I was incessantly swung upward and downward until I thought I should perish in the night. When at length I slept, a carbuncle, which covered the upper false rib of my right side, broke, and from it, at first, there was a scanty black discharge. Luckily, owing to a dose of my father's prescription which I swallowed four times a day, such a copious sweat broke out upon me that it drenched the bed and dripped down from the boards to the floor.

In my twenty-seventh year I was taken with the tertian fever. On the fourth day I was delirious, and on the seventh as well; on that day, also, I began to recover. Gout attacked me when I was at Pavia in my forty-fourth year, and when I was fifty-five I was troubled with daily fevers for forty days, at the crisis of which I was relieved of one hundred and twenty ounces of urine on October 13, 1555. In 1559, the year I returned to Pavia, I was taken with colic pains for two days.

The symptoms of weakness which attended my state of health were varied. To begin with, from my seventh year until I was almost twelve, I used to rise at night and cry out vaguely, and if my mother and my aunt, between whom I used to sleep, had not held me I often should have plunged out of bed. Likewise, my heart was wont to throb violently, but calmed down soon under the pressure of my hand. To this was due the peculiarity of my breathing. Until I was eighteen, if I went out in the wind, particularly a cold wind, I was not able to breathe; but if I held my breath as soon as I became aware of the difficulty, normal respiration was quickly restored. During the same period, from the hour of retirement until midnight I was never warm from my knees down. This led my mother, and others as well, to say that I would not live very long. On some nights, however, when I had warmed up, I became entirely drenched with a sweat so abundant and hot that those who were told of it could scarcely believe it.

When I was twenty-seven, I took double tertian fever, which broke on the seventh day. Later, I had daily fever for forty days when I was fifty-four years old.[3]

In November of my fifty-sixth year, from drinking a mild draught of squill wine, I was taken with dysuria, very acute in form. First I fasted thirty-four hours; later, twenty more. I took some drops of pine gum and cured myself.

It was my custom—and a habit which amazed many—when I had no other excuse for a malady, to seek one, as I have said, from my gout. And for this reason I frequently put myself in the way of conditions likely to induce a certain distress—excepting only that I shunned insomnia as much as I could—because I considered that pleasure consisted in relief following severe pain.[4] If, therefore, I brought on pain, it could easily be allayed. I have discovered, by experience, that I cannot be long without bodily pain, for if once that circumstance arises, a certain mental anguish overcomes me, so grievous that nothing could be more distressing. Bodily pain, or the cause of bodily distress—in which there is no disgrace—is but a minor evil. Accordingly I have hit upon a plan of biting my lips, of twisting my fingers, of pinching the skin of the tender muscles of my left arm until the tears come. Under the protection of this self-chastisement I live without disgracing myself.

I am by nature afraid of high places, even though they are extensive; also, of places where there is any report of mad dogs having been seen.

At times I have been tormented by a tragic passion so heroic[5] that I planned to commit suicide. I suspect that this has happened to others also, although they do not refer to it in their books.

Finally, in my boyhood, I was afflicted for about two years

with indications of cancer. There appeared, by chance, a start upon the left nipple. The swelling was red, dark, hard, and eating. Some swollen veins seemed to remove this toward my young manhood, and in that period a palpitation of the heart—before mentioned—succeeded the varices. From this cancerous growth came blood-blisters, full of blood, and an itching and foulness of the skin; and subsequently I was healed, contrary to all hope of any relief, by a natural sloughing of the mass of diseased skin, although I had removed some of the affections by means of medication.

7.
SPORTS AND EXERCISES

AT A VERY early period in my life, I began to apply myself seriously to the practice of swordsmanship of every class, until, by persistent training, I had acquired some standing even among the most daring. I used to exercise myself with the sword alone, or with sword and shields of various sizes and shapes—oblong, round, large, or small; and I learned to handle as well, dagger, knife, spear, or lance. As a lad, also, with a sword on my hip and a cloak flung dashingly around me, I used to mount a wooden horse.

Another feat I acquired was how to snatch an unsheathed dagger, myself unarmed, from the one who held it. I trained myself by running and jumping, for in these exercises I was reasonably strong. My arms were rather too thin for performances demanding much muscle. In riding, in swimming, and in the use of firearms I had little confidence in myself; I actually feared the discharge of a gun as if it were the wrath of God; for timid by nature, I stimulated courage by an artificial show of bravery, and to that end had myself enrolled in a company of emergency troopers.

By night, contrary even to the decrees of the Duke, I armed myself and went prowling about the cities in which I dwelt. During the day I walked abroad shod with sandals reinforced with lead and weighing about eight pounds. At night I wore a black woolen hood to conceal my features, and put on shoes of sheep-pelt. On many a day, fully armed, I walked for exercise from mid-morning until evening, and often I wandered abroad throughout the night until day broke, dripping with perspiration from the exertion of a round of serenading on my musical instruments.

When making professional calls as a Doctor of Medicine, I went mounted either on horse or mule, or, even more frequently, I went afoot.

From the dinner hour on I am accustomed to wear lighter garments, but when driving I am always more heavily dressed.

8.
MANNER OF LIFE

IT IS MY custom to remain in bed ten hours, and, if I am well and of fair and proper strength, to sleep eight hours; in periods of ill-health I can sleep but four or five hours. I arise at the second hour of the day. If insomnia troubles me, I get up, walk around the bed, and count to a thousand many times. I also diet, cutting down on my food by more than half. At such times I make small use of medication beyond a little poplar ointment or bear's grease or oil of water lilies. With this I anoint seventeen places: the thighs, the soles of my feet, the cervix, the elbows, the wrists, the temples, the regions of the jugular, heart, and liver, and last of all my upper lip. I was especially troubled with early morning wakefulness.

Breakfast was always a lighter meal than dinner. After my fiftieth year I was satisfied in the morning with bread steeped in broth, or even at first, with bread and water, and with those large Cretan grapes called Zibbibos or red raisins. Later in the day I followed a more varied menu, desiring for the midday meal simply an egg-yolk with two ounces of bread or a little more, and a mild draught of sweet wine, or none at all. Or if the day happen to be Friday, or Saturday, I try a small piece of meat with bread and cockle broth or crab soup. I consider nothing better than firm young veal, beaten tender with the back of a butcher knife and pot-roasted without any liquors save its own. In this dish I take great satisfaction; it has a way of drawing its own drippings, than which nothing is better; and thus the meat is far more juicy and much richer than meat roasted on a spit.

For supper I order a dish of beets, a little rice, a salad of endive; but I like even better the wide-leafed spiny sowthistle, or the root of the white endive. I eat more freely of fish than of meat, but only wholesome fresh fish. I love firm meat, the breast of veal or of wild boar roasted, and finely cut with sharp knives. And it seems good to me to eat my meal by the fire. At this repast I delight also in sweet new wine, about six ounces with double, or even more, the amount of water. I find especially good the wings, livers, and giblets of young fowl and pigeons.

I am especially fond of river crabs, because, while my mother carried me, she ate so many of them. Likewise I delight in cockles and oysters. I prefer fish to meat, and eat them with much more benefit than the latter: sole, turbot, flatfish, gudgeon, land-turtles, chub, mullet, or red mullet, roach, sea bream, the merluce or seacod, the spigola or wolf-fish, tilefish, and grayling.

Of the freshwater fish, I prefer pike, carp, perch, and both varieties of sargo. Also, I like loach, squalus,[1] tunny, and herrings, the latter fresh, salted, or best of all, dried. It is surprising that I eat cockles as a very agreeable dish, and turn away from the palatable conger eel, and the mussels as if they were poison, as I do from snails unless they are well cleaned. Fresh water crabs and other crustaceans please me; sea-crabs are too tough, and eels and frogs I find too disgusting, as is the case with fungi.

I take great delight in honey, in cane sugar, dried grapes, ripe grapes, melons—after I learned of their medicinal properties—figs, cherries, peaches, and fruit syrup; nor do these cause me the least distress, even at my present age. Above all, I find olive oil delicious, mixed with salt and ripe olives. Garlic does me good; and bitter rue has always seemed to have special properties, both when I was a boy and since I am grown, of protecting my health and, besides, serving as an antidote against all poisons. By experiment I have found *absinthium Romanum*, or wormwood, beneficial.

I was never immoderately addicted to venery, nor have I been harmed much by excesses in this respect; now, however, it plainly results in abdominal nervousness.

The white meat of the smaller fishes, when they are fresh and tender and cooked on the grill, I enjoy and find beneficial. Nor do I turn up my nose at a good sheep-cheese. Above all eatables, I prefer a carp of from three to seven pounds, but of the choicest flesh. From the large fishes I remove the head and the entrails, and from the little ones the backbone and the tail. The head is always cooked in boiling water, and, in the case of the larger fishes, all remaining parts either fried or broiled on the gridiron. Small fishes may be fried until tender, or boiled not too much.

Of the flesh of animals, the white meats are the better. The heart, liver, and kidneys are tougher than the lung, which is tender, although the lower parts of the latter offer little nourishment. In some animals the red meats, excepting the heart, are tender, the white but indifferent, except the testes, which are tender. The bluish parts are less easily digested.

There are seven principal genera of things: air, sleep, exercise, food, drink, medication, and preservative.

There are fifteen species: air, sleep, exercise, bread, meat, milk, eggs, fish, oil, salt, water, figs, bitter-rue, grapes, and strong onions.

Preparatives are also fifteen: fire, ashes, the bath, water, the stew-pan, the frying-pan, the spit, the gridiron, the pestle, the blade of the knife and back of it, the grater, parsley, rosemary, and laurel.

The exercises are turning the mill-wheel, walking, horseback riding, playing ball,[2] carriage driving, fencing, riding, the saddle, sailing, the furbishing of plate,[3] massage, or bathing—fifteen![4] I have reduced the whole to a system as is the fashion in matters of theology, with much profound meditation and brilliant reasoning.[5] For without this illuminating logic, certain things, which are actually most clear, would seem not quite so evident to you!

There are five things which may be partaken of in proper manner by all cxcept old men: bread, fish, cheese, wine, and water. Two medicines are mastic and coriander; these should be highly sweetened. Two spices are saffron and salt, and this last is also an element. Four things must be used in moderation, for they are highly nourishing foods: meats, yolks of eggs, red raisins, and oil; the last is an unknown element which, when subjected to fire, corresponds to the properties of the stars.

9.

A MEDITATION ON THE PERPETUATION OF MY NAME

VOWING to perpetuate my name, I made a plan for this purpose as soon as I was able to orient myself. For I understood, without any doubt, that life is twofold: the material existence common to the beasts and the plants, and that existence which is peculiar to a man eager for glory and high endeavor. In the former, I realized that nature had failed me, that my desire had been left ungranted; as for the latter, I knew there would be nothing by reason of which I would dare to hope—neither resources, nor power, nor firm health, nor strength, nor family, nor any special devotion to labor. I did not have a wide knowledge of the Latin tongue, nor friends, nor anything from my parents except an endowment of misery and scorn.

After a few years I was inspired by a dream to a hope of attaining this second way of life—the way of fame. Only I did not clearly see how, except in so far as I was helped, as it were, by a miracle to the understanding of the Latin tongue. But in truth I was recalled by my sane reasoning from any great aspiration toward such fame, perceiving that nothing was emptier than that hope, not to mention my simple resolve.

"How," said I, "will you write what will be read? And what remarkable facts do you know that readers care for? In what style will you write, or with what choice of diction, so that you may hold the attention of those readers? Can it be that any would read? May it not be that the course of ages will see such a constant accumulation of writings that those early books may be scorned, not to say, neglected? Will they endure for even a few years? How many—a hundred, a thousand, ten thousand? Show me a case; is there one such book among thousands?

"And since all things will come to naught—even as there was a beginning will there be an end—even though, as the academic philosophers want to believe, the world may again be renewed, does it make any difference whether the end is after ten days, or after ten times countless thousands of years? None either way, and it is all one in eternity!

"Meanwhile, will you torment yourself with hope, will you be tortured with fear, will you be exhausted by strivings? Whatever of sweet life is left, you will lose—oh excellent idea!"

Yet did not Cæsar, Alexander, Hannibal, Scipio, Curtius, and Herostratus[1] prefer this hope of enduring fame before all others, even at the risk of infamy, the price of torture, and the cost of very life itself?

So it is; and though it be nothing, nevertheless they succeeded in large measure in realizing their ambitions. They did not glance at that philosophizing of the sages, much less adopted any of it; but they exerted themselves toward this one end—fame.

And yet again who denies that it was absolute folly? As such it is recognized by the judgment of Horace himself in Ode 29 of Book III: *Tyrrhena regum progenies:*

> Ille potens sui
> Lætusque deget: cui licet in diem
> Dixisse, vixi, cras vel atra
> Nube polum, pater occupato,
> Vel sole puro; non tamen irritum
> Quodcumque retro est efficiet, neque
> Diffinget, infectumque reddet,
> Quod fugiens semel hora vexit.[2]

He had come to the following conclusion, moreover, somewhat earlier saying: *Quod adest memento componere æquus*; that is to say, "You may profit more by accepting the present, than by making arrangements for the future."

Cæsar, Hannibal, and Alexander had this design, to advance to distinction their own names at the cost of their lives, and at the same time the sacrifice of their families, their followers, and even of their city or their state, making, in the meantime, the most of their own positions. Suppose, then, that in this manner they achieved fame. To what end? Sulla destroyed the institutions all his predecessors had labored to build; indeed, whatever was before his day, even things most admirable. All of his successors caused their families and their friends to perish. The whole Julian house was completely swept away by Commodus, during the period when that prince, adulterous and false at every turn, let his suspicions fall upon any legitimate descendant of that family. And in the same way he ruined his country. For where now is the Roman Empire? Absurdly and strangely enough—in Germany!

How much better would it not have been for the illustrious Julian house to have survived—the race of Æneas; for the Romans to have been the masters of the world than for hollow masks and stuffed effigies of men to be decorated with these empty names! And so, if the spirit is imperishable, what is the use of vain names; if it is perishable, what do they avail? If a whole generation will pass away, will not all these glories come to an end, will they not perish, even as the hares and rabbits of the fields?

It is scarcely surprising, therefore, that I, urged on by love of fame, seek it; what *is* surprising is that I can still seek it, realizing all these matters which I have just considered. But be that as it may, an unshakable ambition remains. It was, I grant, a fatuous purpose in Cæsar and these others; yet my desire is for renown, so many things to the contrary, so many obstacles in my way; and it is a desire not so much foolish as stubbornly fixed.

Yet have I never longed for praise and honors; indeed, quite the contrary, I have spurned them, wishing it to be known only that I had lived, and having no concern that it be known what manner of man I was.[3]

As for my descendants, I know how fraught with uncertainty this hope for fame may be, and I realize how little we may foresee its consequences. Therefore, I have lived my life as best I might; and in some hope of the future, I have scorned the present. If I must excuse my present manner of life let me say that I now continue to exist as well as I can. For this course seems but honorable; and even if any hope I have for fame should fail me, my ambition is worthy of praise, inasmuch as longing for renown is but natural.

10.

CONCERNING MY COURSE OF LIFE

GUIDED by the foregoing philosophy, therefore, I determined upon a course of life for myself, not such as I would have, but such as I could. Nor did I choose perhaps exactly what I should, but what I deemed would be better. And in this my purpose was not single or constant, since every course is full of danger and of hardship, and far from perfect. Accordingly I acted as seemed advantageous when each occasion arose; thus it has come about that I—by comparison with others—am considered an opportunist, and even, as I have said, by no means steadfast of purpose. For such men as have no sure procedure in life must, perforce, try many plans, and make progress through devious ways. And I, to the end that a certain continuity of aim might be gained, did not permit riches, nor ease, nor honors, nor magistrates, nor power, nor, in truth, this very ambition, nor good or evil vicissitudes, nor rivals, nor the exigency of the times, nor my own ignorance, to stand in my way; nor yet the fact that I had, obviously, no qualifications for living aimfully. The very knowledge of astrology which I had at the time was, moreover, prejudicial, for it seemed to show me, and all my acquaintances declared, that I would not pass my fortieth year—that I surely would never live to be forty-five.

Meanwhile, partly out of necessity, partly tempted by pleasure, I continued to transgress daily, even while I was deliberating upon how best to live my life. Neglecting my own best interest for a false hope, I deviated from my purpose of taking counsel concerning these interests, and I sinned more than once in deed. Finally it came to this, that the very year in which it was believed the end of my life was at hand, brought with it a beginning of living—and that was my forty-third year. That was the moment when, induced by my age, by my disposition, by the anxieties of the past and the opportunity of the present, I made a beginning, turning away from pleasure.

In the morning, if I were teaching, as I was first at Milan, and later off and on for many years at Pavia, I delivered my lecture. That over, I went walking in the shade beyond the city walls, and later lunched, and enjoyed some music. In the afternoon, I went fishing near a grove or in a wood not far from the city. While there I also studied and wrote, and in the evening returned home. This period lasted six years; but alas—*Fulsere quondam candidi tibi soles*,[1] as the poet says.

Thereafter I entered upon a long and honorable career. But away with honors and gain, together with vain displays and unseasonable delights! I ruined myself! I perished! Difficulties and evils increased like the shadow of the yew-tree, as the saying goes. Now no solace remained, except the deadly way of destruction. But therein no blessing can be found; for otherwise despots, who are actually farthest from bliss, would be the most blessed. Just as the bull, raging ahead with fixed eyes, sweeping on in headlong career, must dash himself to ruin, so I sped precipitously to my downfall. For in the midst of things, and before these present events, the great disaster overtook my elder son.

Certain members of the Senate—I think they would scarcely wish it to be known about themselves—have actually confessed that they condemned him with the hope that I should die of grief, or go mad; and how near I came to either end, the gods alone know—and this I shall set forth in the proper place; but their purpose failed. I wish my reader to know—and for this purpose I digress—what were the iniquitous *mores*—how ungodly the times! For it is a truth that I never offended one of these men, not even by my shadow. I planned whatsoever defense I was capable of in behalf of my son; but who could have prevailed against the festered disaffection of certain of those senators?

I was shocked by the memory of my son's despair, in consternation over the impending dangers, exhausted by all that had gone before, and fearful of what was to come. In such a state, however, I made my plea, calling attention to the high character and the equity of the Senate, and mentioning cases where mercy had been shown. I recalled the lenity, and at the same time, the act of Giampietro Solario, the notary, who, when he had charged his illegitimate son with attempting to poison two legitimate sisters for the sake of becoming the sole heir to his father's property, considered that the boy was sufficiently punished by being condemned to the galleys.

Augustus was praised for this question when in an examination he asked, "At least you have not killed your father?" What cruelty, then, to destroy an innocent and aged father through the person of his son! If the father is taken into kindly consideration in the case of a man condemned to the arena, how much more should he be favored in the case of any other condemnation? Of what avail are the merits of humankind if so exceptional a virtue as innocence is so disastrously involved?

And is it not worse for the father to be involved in the punishment of his son than in his own destruction? If I am killed, one man perishes—one about to die without another descendant; if you kill the son, you cut off all hope of succession. Imagine each one of all mankind pleading with you for his son, for whom he feels responsible, though he may be a youth of hasty temper, struggling with difficulties, overtaken by the basest dishonor and deceived in a dowerless wife, having married against a father's wish and knowledge, a worthless, shameless woman—what would you? Is this not every man's plea? Does not everyone understand? No one is so bitter an enemy of me or of my son who would not voluntarily grant life to him whose death would move even the iron gods of the underworld to pity!

Although I brought forward these arguments and their like, it availed naught except in so far as it was decreed by the court that if I should be able to come to terms with those who had brought the charge against him, his life would be spared. But the very indiscretion of my son forbade this; for he had boasted of riches which I did not possess, and the accusers tried to exact what did not exist.

But no more of this!

From my early youth I persistently held to this purpose—that I should make it my duty to care for human life. The study of medicine seemed to point more clearly to such a career than did the study of law, as being more appropriate to the end I had in view, and as of more common concern to all the world in every age. I deemed medicine a profession of sincerer character than law, and a pursuit relying rather upon reason and nature's everlasting law, than upon the opinions of men.

Accordingly, I embraced this pursuit and not jurisprudence. Thereupon, deliberately, I not only rejected the advances of friends engaged in the law, contemning riches, power, and honors, but even shunned these influences. My father actually wept in my presence when he learned that I had given over jurisprudence to follow the study of philosophy, and felt deeply grieved that I would not apply myself to his same interests. He considered jurisprudence a more ennobling discipline—repeatedly he quoted Aristotle on this point—and a profession better adapted to the acquisition of wealth and influence, and to the improvement of the family position. He realized that his office of lecturing in the law schools of the city, together with the honorarium of a hundred crowns which he had enjoyed for so many years, would not, as he had hoped, fall to me, but that another would succeed him in his post. Nor would that commentary of his ever be published, which I was to annotate. For not long before this there had dawned a faint hope that he might achieve some renown as the critic of *The Commentaries of John, Bishop of Canterbury, on Optics and Perspective.* The following couplet had even been published for this work:

> Hoc Cardano viro gaudet domus: omnia novit
> Unus; habent nullum sæcula nostra parem.[2]

This might rather be considered in the manner of prophecy for one who was then about to set out upon his life's labors, than as applying to my father himself, who, beyond the law—which, I understand, he practiced with extraordinary brilliance—had mastered only the elements of mathematics; he was in no wise given to original thinking, nor had he availed himself of the resources of the Greek language. This situation came about, in his case, more because of his many-sided interests, and his inconstancy of purpose, than because he was not naturally gifted, or because of sloth or faulty judgment; for he was subject to none of these defects. However, because my will was firmly set to my purpose on account of the reasons which I have already advanced, together with other motives, I was not moved by my father's advices, especially since I saw that he, although he had met with practically no reverses, had succeeded but indifferently.

11.
PRUDENCE

IT IS SOMETIMES better to persist in a course elected, even when the course has not been too well considered, than to shift, in an effort to make a perfect choice, from course to course, albeit an intensity of zeal may importune, or the usual inconstant ebb and flow of mundane affairs may urge the change. Accordingly, when I had made a decision for myself in a most difficult matter, to the end that my action should be governed by wisdom as well as by my own preferences, I realized then that not only other things, but also this very business of remaining firm in my choice was made easier. Of first consideration are the various ends of action which lead each man to choose the one toward which his taste is most inclined. There are so many fashions in careers, so many chances, so many interests and so many opportunities, that no one may justly censure me unless it be one who confesses that he lives more in accordance with my plans and purpose than I myself—and this can in no wise be.

All granted up to this point, the next question which arises concerns the best and most legitimate method of succeeding in a chosen career, or—of even greater moment, to my thinking—what is the most expedient? Having, then, once attained the goal in view, how to retain it, and how, finally, to draw just advantage from the end achieved, is important.

Now I have ever attested my insignificant powers in *εὐβουλία* or *φρόνησις*. If these terms signify merely *prudentia*, it is, nevertheless, the same as if we should say *humana prudentia*, for nothing which we know, except man, possesses this good counsel. The deities have a greater gift—that is to say, direct knowledge or intuition, not a quality of other existing creatures. *Intuitum* is not an expression identical with the harpocratic quality, since it differs in character from *foreknowledge* in that it partakes of the nature of *prudentia.* To some it presents one appearance, to others another, as men are wont to estimate all things by their own slant.

Truly of this sort of human refinement and sagacity I know I possess and have possessed very little, and what there was, indeed, those habits I have already mentioned have corrupted.

12.

DEBATING AND LECTURING

AS A LECTURER and debater, I was much more earnest and accurate than in exercising prudence. While at the University of Bologna, I usually lectured extempore. This habit always undermined the confidence of those with whom I argued.

A three-day debate was instituted at Pavia with Camuzio, to be held in public before the Senate. My opponent was silenced, on the first day, in the first proposition, even in the judgment of all my rivals who were present. Certain memorials of this same event are graven in letters on the monument to Camuzio[1]: "This then, in truth, was known to all, that they debated not for a mere refutation of argument, but with a power which seemed unassailable." And I believe the memory of that debate lives to this day.

Branda, who was, as I have stated, my preceptor, attributed my powers to art and talent; my rivals said I was possessed; others, conjecturing much more accurately, claimed my powers were due to a certain superiority and perfection in the line of reasoning. For neither at Milan, nor at Pavia, nor in Bologna, nor in France or Germany, have I ever found a man who could successfully controvert or dispute me within the last twenty-three years. Yet I do not vaunt my powers on this account, for I think that, had I been made of stone, the same things would have come to pass. It is the result of the lack of clear thinking on the part of those who would challenge me, and no more a special dispensation to my own nature or to my own distinction, than it can be counted glorious for the cuttlefish to eject the shadows of its inky humor about the dolphin and force it to flee; that is merely the result of being born a cuttlefish.

Angelo Candiano in the presence of many scholars, when he had brought forward his proofs and I had answered, and wished him to take up the rebuttal in his turn, did not blush to say, "I stated that I wished to bring forward a proof, not to respond to your arguments." Now he was a physician of great learning, who had filled high offices at the court of the Duke of Milan[2] and at the court of the Queen of Hungary in Belgium.[3] He was, besides, an authority in his profession, and, if it is of any consequence, very rich. Whereupon, when I confessed my inexperience and simplicity, many insisted, saying, on the one hand, "We know that you are speaking falsely and that you are experienced"; and, on the other hand, "We ask, since we cannot clearly see, why you have recourse to that expression about 'inexperience,' especially strange on the lips of a man who has asseverated so many times that he never lies. As to your inimitable style of teaching, let us say that what is in the positive—as the grammarians say—ceases to arouse our admiration when the superlative itself becomes the common thing. That these present have made no move to express a desire to see a demonstration of it, is not significant. The sun does not cease to be because masses of clouds hide it from view. Nor should you be troubled because, though so many bright lamps of learning burn within your private chambers, those without your doors appear to refuse to see the radiance. There need be no fear, moreover, that an endowment so divine may come to naught. Flowers still adore the rising sun even while the Garamantes curse it.[4] Over all not only is Divine Providence enthroned, but there shines, as well, an eternal light of wisdom."

Not only have I always distinguished myself, moreover, in this gift of easy delivery, but I have instructed others therein. Excellent as I may have appeared in these respects, I possessed neither grace in my manner of speech nor talent for making a clever conclusion. The result was that though you deemed me more than usually endowed on the one hand, on the other you would find me lacking. Finally, in disputation I was so exceptionally keen that all marveled at my exemplary skill, and avoided challenging me. For a long time, consequently, I lived free from the onus of debate, but not longer than the time when my opponents had unexpectedly seen two occasions of my skill.

The first of these occasions was at Pavia. Branda Porro, formerly my master in philosophy, had interrupted the course of an informal debate which I was holding with Camuzio in philosophy; for, as I have said, they often dragged me into this field, thinking that in the province of medicine they had no further hope of forensic laurels. Branda was citing Aristotle as an authority, whereupon I, when he had quoted the words, said, "Take care; there is a *non* after *album* which you have forgotten to include, and which contradicts your proof." Branda exclaimed loudly, "You don't mean it!" I, clearing my throat of the phlegm which constantly collected therein, gently maintained my opposition, until he, thoroughly angered, sent for the codex, and upon my request, ordered the text to be put into my hands. I read as it was written therein; but he, suspecting that I was falsely emending the text, snatched the book from my hands, cried out that I wanted to cheat the audience, and himself began to read. As he came to the word in question, he read it and was silenced. All present were amazed and stared at me in wonder.

It happened also during these same days that Branda went to Milan. The whole incident had been written to the Senate at Milan. The senators asked if the story were true. Branda, a sincere and honest man said, "Surely it is true; only too true; I believe I was drunk on that occasion." The senators, smiling, had nothing to say.

The second occasion on which I was challenged was at Bologna, by Fracanziano, Professor of the Practice of Medicine. He was engaged in a discussion about the passage of gall to the stomach, and was quoting from a Greek authority in the presence of the whole academy, for an anatomical dissection was then in progress. I said that *οὐ* was lacking in his citation. Whereupon it was not he, for a truth, as I quietly defended my correction, but the students, who cried out that some one should send for the codex. Fracanziano gladly sent, and it was brought at once. He read, and found that it was as I had said, to a hair. He was silenced, amazed, and filled with admiration; the students, who had literally dragged me forcibly to the place, marveled even more.

From that day, the professor actually fled from any occasion for meeting me; and he even warned his servants that they indicate whenever they saw me coming, so that he might avoid an encounter in the street. And when once, for a joke, the medical students had fetched me into his presence when he was busily engaged in a lesson in anatomy, he hurried away in such confusion that he stepped on his cloak, and fell headlong. All who witnessed this were bewildered at his action; and he himself shortly afterwards resigned as a professor, for he was at that time a man well along in years.

13.
CUSTOMS, VICES, AND ERRORS

AN ACCOUNT of this nature is by its own character a most difficult thing to write, and so much the more for me as I reflect that those who have been wont to read the autobiographical books of writers are not used to seeing such a straightforward narrative set down as I purpose to publish.

Some have committed themselves to writing as they think they ought to be, like Antoninus; others have, indeed, given true accounts, but with all their shortcomings carefully suppressed, as did Josephus. But I prefer to do service to truth, though well aware that he who transgresses the conventions cannot offer the same excuses as suffice for other mistakes. Yet who constrains me? Shall I not be, therefore, like that leper who, of the ten healed, alone returned to acknowledge the Lord?

By this reasoning physicians and astrologers find the origins of our moral natures in our innate qualities, and of our voluntary acts in education, interests, and conversations. These all are present in every man, but peculiarly adapted to each appropriate epoch of his life, yet with variations, nevertheless, even in similar instances. It is necessary, consequently, to exercise choice, and to select with care from all these influences one principle whereby we may get understanding. For myself, as far as I could judge, the well known *γνῶθι σαυτόν*[1] seemed the best guide.

My nature, as a result, has ever been manifest to me: high-tempered, straightforward, and devoted to the pleasures of passion. From these beginnings, as it were, have issued bitterness, contentious obstinacy, lack of amenity, hasty judgment, anger, and an intense desire for revenge—to say nothing of headstrong will; that which many damn, by word at least, was my delight. "At vindicta bonum vita iucundius ipsa."[2] On the whole I was not inclined to deviate from the ways of rectitude even though it is commonly said, "Natura nostra prona est ad malum."[3]

Yet I am a truthful man; I am mindful of benefits conferred, attached to my own people, a lover of justice, and a contemner of money. Zeal for undying fame has captured my devotion, and rendered me wont to despise mediocrity, to say nothing of petty concerns. Aware nevertheless, how greatly very insignificant matters affect any circumstance from its outset, I am accustomed to minimize no occasions. By nature I am prone to every vice and every evil save ambition; I recognize my shortcomings as well as anyone.

For the rest, because of my veneration for God, and because I clearly recognize how vain all these things are, I deliberately let pass many an offered opportunity for revenge. Timid of spirit, I am cold of heart, warm of brain, and given to never-ending meditation; I ponder over ideas, many and weighty, and even over things which can never come to pass. I am able to admit two distinct trains of thought to my mind at the same time.

Any who cast aspersions upon the praises I have enjoyed by intimating that I am boastful and extravagant, accuse me of faults of others, for those sins are not mine. I resent such, and defend myself; I attack no one. Why then do I trouble to make this examination of myself when I have given testimony so many times of the emptiness of life? My excuse is the praise spoken by some, who think that a man who has attained so much distinction does not have his shortcomings.

I have accustomed my features always to assume an expression quite contrary to my feelings: thus I am able to feign outwardly, yet within know nothing of dissimulation. This habit is easy if compared to the practice of hoping for nothing, which I have bent my efforts toward acquiring for fifteen successive years, and have at last succeeded. And now, trained to pretenses of a sort, I sometimes go forth clad in rags, but just as often elegantly dressed; sometimes I am taciturn, and sometimes talkative; sometimes gay, sometimes sad. From these moods all things acquire double aspects.

In youth I spent little and only occasional care upon the appearance of my head because of my eagerness to be at things I was more interested in. My movements are irregular, now quick, now slow. At home I go about with my legs bare to the ankles.

I am lacking in reverence and have a far from continent tongue; and I have such a quick temper that I am often shamed and embarrassed thereby. Though I may be led to repent, I have, nevertheless, paid heavy penalties, as I have elsewhere stated, for my sins; even so I atoned for the debaucheries of that Sardanapalian life I led the year I was rector of the University at Padua. Yet wisely and patiently to have carried and corrected one's fault, is praise even in disgrace, and virtue even in sin. Necessity may pardon me for speaking thus in my own praise; yet even if I were inclined to pass over in silence the gifts of God toward me, I should be an ingrate indeed to complain of the loss which I suffered without acknowledging the kind of life I was leading at the time. Frankness is, moreover, the simpler course inasmuch as my personal affairs are not as highly esteemed as men commonly value their own interests—vain, empty affairs like those great clouds seen in the wake of the sunset which are meaningless and soon pass away. If anyone would like to pass unprejudiced judgment on these actions and to reflect how easy it is to yield, he will understand in what spirit of mind, and urged by what necessity, or by what occasion I acted, and with what great grief I was afflicted through those deeds of my youth.

It is a fact that some, without scruple, are guilty of faults much worse than those; who, far from confessing them publicly, are silent even in private. They are people who are not mindful of benefits conferred upon them—who entirely forget that favors have ever been done them. Perhaps these people will be somewhat more charitable in judging me.

But let us get on. This I recognize as unique and outstanding among my faults—the habit, which I persist in, of preferring to say above all things what I know to be displeasing to the ears of my hearers. I am aware of this, yet I keep it up wilfully, in no way ignorant of how many enemies it makes for me. So strongly are our natures fettered to long-standing habits! Yet I avoid this practice in the presence of my benefactors and of my superiors. It is enough not to fawn upon these, or at least not to flatter them.

I used to be just as immoderate in living when I well knew what course was most expedient to follow, and what I ought to do; scarcely another man could be found so obstinate in a fault of the sort.

I am alone as much as I can be, although I know that this way of living has been condemned by Aristotle. For he said, "Homo solitarius aut bestia aut deus." But for this I have rendered an excuse.[4]

By a similar foolishness, and with no small loss to myself, I retained those domestics whom I knew to be utterly useless to me, or even a cause for shame. I become the owner of all sorts of little animals that get attached to me: kids, lambs, hares, rabbits, and storks. They litter up the whole house. I have troubled myself with the poverty of friends, especially my faithful friends.

And I have committed many, nay, numberless blunders, wherever I wished to mingle with my fellows; whatever these were, whether great or small, timely or untimely, I was conscious of them; and I even went so far as to wound those whom it had been my intention to praise. Among these was Aimar de Ranconet,[5] a president of the Parliament of Paris, a most devoted scholar and a Frenchman. On this occasion I blundered, almost unavoidably, not solely because of lack of deliberation, and an ignorance of foreign manners and customs, but because I did not duly regard certain of those conventions which I learned about long afterwards, and with which cultivated men, for the most part, are acquainted.

In deliberating, I am too precipitate, and therefore frequently make rash decisions. In any sort of transaction, I am impatient of delay. My rivals, observing that I was not easily taken in, if I had time to think, did their best to hurry me on. I detected them plainly enough, since I am on my guard against my competitors, and deem that I can justly hold them as my enemies. But if I had not been wont never to regret any action I voluntarily undertook, even something that came to a bad end, I should have lived most unhappily.

Truly the cause of a great part of my misery was the stupidity of my sons, connected as it was with actual shame, the folly of my kinsfolk, and the jealousy existing among them, which was a vice peculiar to our family. Indeed, it is a fault common to many a group!

From my youth, I was immoderately given to gambling; and in this way I became known to Francesco Sforza, a prince of Milan, and made many friendships for myself among the nobles. And since for many years—almost forty—I applied myself assiduously to gambling, it is not easy to say how greatly the status of my private affairs suffered, and with nothing to show for it. The dice turned out to be far worse, and once my sons were instructed in the attractions of games of chance, our home too frequently was thrown open to gamblers. For this gaming habit I have naught but a worthless excuse to offer—the poverty of my early life, a certain shrewdness in hazards, and something of skill in play.

This is mortal frailty, then, but some will not admit it; others are even intolerant of it. Are they better, or wiser?

What if one should address a word to the kings of the earth

and say, "Not one of you but eats lice, flies, bugs, worms, fleas—nay the very filth of your servants!" With what an attitude would they listen to such statements, though they be truths? What is this complacency then but an ignoring of conditions, a pretense of not being aware of what we know exists, or a will to set aside a fact by force? And so it is with our sins and everything else foul, vain, confused and untrue in our lives. On a rotting tree are rotten apples! It is nothing new that I proclaim; I merely lay bare the truth.

14.
VIRTUES AND CONSTANCY

ALTHOUGH there are many points on which men are confused, in none are they more deceived than in their conception of firmness of character or constancy.

This confusion originates in the very nature of constancy, a characteristic which in one man may truly be an evidence of divine endowment; in another a trait, rather, significant of a small, unthinking mind. It is a trait which some one derided in Diogenes because under the summer sun he rolled himself in burning sands; in winter with bared body he embraced columns of ice. On the other hand, it was the high virtue of Bragadino,[1] the Patrician of Venice, to endure such trials as no man, not even one of the host of arrogant conquerors, was eager to administer, yet such trials as are worthy of an immortal name; for he was flayed alive. If it was nothing short of superhuman to endure such martyrdom, surely it was compatible with human longing for an immortality of fame to have willed to endure.

And albeit a man may shine with serener soul in the midst of reverses, not infrequently he is given an opportunity in the midst of prosperity to show himself worthy of admiration. Moreover, even if occasions to display their steadfastness may fail some men, not on this score ought they be deemed less constant. Since, then, it happens that we err in so many respects in the name of constancy, mere endurance ought not to be thought a matter for glory, any more than lack of occasion to demonstrate our firmness of character ought to be thought reason for blame; nor should anyone hold us blameworthy for what nature has seen fit to deny us.

Neither do I defend myself on the ground that opportunities have in some degree been lacking. No one was ever so bitter against me, no judge so unjust, as not to admire my patience in adversity, and my restraint in prosperity, rather than criticize it, whether that virtue resided in my power to despise the attractions of joy, or in my courage to bear hardships.

I name, among the trials of my life, the indulgence of the flesh, distracting entertainments, disease, impotency, the disparagements of my rivals, issues none too felicitous, lawsuits, attacks, the threats of the powerful, the suspicion of certain men, the distractions of a family, the failure of many projects; and finally, the contrary advice either of true friends, or of men in the guise of friends, and the hazards which beset me on account of my many unorthodox views.

Notwithstanding, whatsoever my good fortune, or however many the happy issues that attended me, I have never modified my carriage, nor become more bitter or inordinately ambitious, nor more impatient, nor contemptuous of the poor, nor forgetful of my old friends, nor more abrupt in social intercourse, nor more boastful. My good fortune did not tempt me to adopt a more luxurious mode of dress than I felt was demanded by my professional rank, except in so far as it was necessary to replace with better the out-of-fashion garments my poverty had compelled me, from the beginning, to wear.

In misfortune, however, I am by no means steadfast, since certain afflictions which I was forced to meet were more than I could bear. For these exigencies I overcame nature by a scheme of my own. In the moments when my spirit was afflicted with the most insupportable grief, I would strike my thighs with rods, or bite my left arm sharply and quickly. Often

I fasted. I was relieved much by weeping, if weep I could, but very often I was unable.

Moreover, I combated my sorrow with reason saying, "Nothing new has come to pass; my times have nearly changed and have hurried on apace. Might it not have been possible to escape forever this association with sorrow? Yet if I have been cheated for a few years, what portion of time is that measured by eternity? I shall have lost a little, that is all, in the event that I have not long to live; but if a life lengthened by coming years will appear to be my allotted span, perhaps many opportunities may come whereby I may assuage my grief and of it make an eternal triumph. And yet, what if it had never come to pass! Truly I am not equal to this overpowering grief!" And yet, as I shall hereinafter show, I was consoled by a manifest miracle.[2]

To the duties of life I am exceptionally faithful, and particularly in the writing of my books, to such an extent that even though the most attractive opportunities have been offered, I have not abandoned my undertaking, but continued to adhere to my original purpose.[3] I had observed that my father's habit of relinquishing one aim in life for another had been a mighty obstacle to his success.

I do not think anyone would criticize me because, when I was appointed to the Accademia degli Affidati at Pavia[4] in which there were not a few Princes and Cardinals of highest rank, causing me to hold back somewhat out of sheer pusillanimity, I neither disdained the offer by remaining away, nor did I excuse myself from the duties of my office. To be sure, when these men of rank, in full regalia, were presented to the King, I kept in the background, saying to myself that pomp of this sort in no manner became me.

Concerning goodness, moreover, I can say nothing other than Horace:

Virtus est vitium fugere.[5]

I have never broken a friendship; neither, if the relationship happened to be discontinued, have I divulged the secrets of my erstwhile friends, nor held those secrets against them. I have written nothing inconsistent with my principles. Aristotle was somewhat inclined to sin in that sort of thing, and Galen descended so far as to become involved in a disgraceful contention. In this matter I yield to Plato alone. Vesalius, a man of decent restraint, gave proof of his attitude toward malignant controversy, for when he was stirred by Corti to engage in some insignificant dispute, he was not willing to mention his opponent's name.

In spite of the fact that I was envious of Corti's[6] great love of letters—although not his erudition—it happened that when the board, upon his withdrawal from Pisa, inquired of him whether I could qualify as a successor to his post, he replied, "No one would be better!" Now I had been accused by Corti of the theft of a ring which I had retained as a pledge of money which he had made without a witness; and although the members of the board knew that we were not entirely reconciled on this score, they gave me the office of lecturer on his recommendation.

These I deem ought to be accounted among my virtues—that I have never told a falsehood from my youth,[7] that I have borne poverty and disaster and all the pricks of adverse circumstances patiently; that no one has ever been able justly to charge me with ingratitude. But enough of this!

15.
CONCERNING MY FRIENDS AND PATRONS

FIRST AMONG the friends of my youth I count Ambrogio Varadei; together we played many a game of chess, and diverted ourselves with music or similar amusements. After him I name Prospero Marinone[1] of Pavia and Ottaviano Scotto[2] of Milan, who often helped me to a loan of money. Gasparo Gallarati was another.

In the town of Sacco, I formed close friendships with Giammaria Morosini, a nobleman of Venice, and with Paolo Lirici, a pharmacist. After my return to Milan I found a patron in Filippo Archinto, Archbishop of Milan, through whom I became known to Lodovico Maggi whose aid I sorely needed and who gladly helped me.

Among other friends I must name Girolamo Guerrino, a jeweller, from whom I learned many mysterious stories which I have rewritten in my books but with no idea of plagiarizing. Through him also I became involved in relations with Francesco Belloto, a Florentine.

Francesco Croce, the jurisconsult, a noted man of high character and a skilled mathematician, rendered me valuable service in my case with the College of Physicians at Milan. Through the efforts of Donato Lanza, a druggist,[3] I was introduced to Francesco Sfondrato, a Senator of Cremona,[4] who afterwards became Cardinal. Sfondrato recommended me to a criminal judge, also a native of Cremona, Giambattista Speziano, a man of learning and of singular moral quality. He it was who made me known to the great Alfonso d'Avalos, Governor of the Province of Milan and Commander of the

Emperor's forces in the State. By the patronage of Sfondrato, also, I obtained the chair of medicine at the University of Pavia.

At Pavia I was received into the friendship of Andrea Alciati,[5] that most celebrated of jurists, and a great teacher of oratory. His kinsman, Francesco Alciati,[6] who is now Cardinal, joined in this cordial relation. Two other cardinals, besides, I must include in my list of patrons: Giovanni Morone,[7] and Pietro Donato Cesio.[8] Under the protection of these three Mæcenates my present circumstances stand assured. Together with these I must name a fourth, Cristoforo Mandruzio of Trent. He is of a most illustrious family of nobility, and second to no one in his favors toward me and in his liberality toward all men.

After acknowledging my indebtedness to these notables, let me return to those good friends among my equals. My association with Panezio Benvenuto of Arezzo, the best of men, gleams more richly than any gold, as it seems, an attachment enduring by the very virtue of its own noble nature. Again, there was the venerable Taddeo Massa, at Rome, a prelate, and a man of unique wisdom and probity. Long before this, however, I had secured the friendship of Giovanni Meone, from the Privy Council of the Governor of the Province, Fernando Gonzaga,[9] who was also Commander of the military forces of the Emperor.

It would be too long a story to go into the details of my relations with Carlo Borromeo[10] and with Marc Antonio Amulio of Venice, both cardinals of the most exceptional character, and with a great number of other good men. Certainly it was by the effort and influence of Borromeo and Alciati, when I came to Bologna appointed to lecture in Medicine there, that I obtained the favor of the whole illustrious senatorial assembly; for they are all patricians extraordinarily courteous, cultivated, experienced, and brilliant.

Among the members of the medical profession I had two

intimate friends, both men of blameless life, and of by no means mediocre erudition, and both natives of Modena: Camillo Montagnana and Aurelio Stagni. Besides these I must name Melchiorre della Valle of Milan and Tommaso Iseo of Brescia whom I sought after with a good-will unusual for me, but nothing came of the relation except a rather tense animosity.

My friends among the high officials of the English Court were Sir John Cheke,[11] childhood tutor to King Edward VI, and Claude Lavalle, who was the ambassador of the French King to England, and Prince of Bois-Dauphin.

Among my own fellow-citizens I owe not a little to Lodovico Taverna, Prefect of Milan, and a very sagacious man.

I highly esteemed, among my professional associates, Francesco Vicomercato, a Milanese, and a professor of Philosophy; and Andrea Vesalius, the foremost exponent of his day of the science of anatomy.

Two friends of my father, known from childhood, I continued to cultivate: Agostino Lavizzario, clerk of petitions of the Senate of Como, and Galeazzo del Rosso, a blacksmith, whom I have frequently mentioned.

There was also Francesco Buonafede, a doctor of Padua, referred to on another occasion.[12]

I must pass over a description of many learned men and friends, because, even without formal mention, they are known for their erudition throughout the world. Yet, in order that you may actually know, since I have witnessed their favor to me, that I am not forgetful of them to whom as far as I am able, I am about to give lasting recognition by mention of their names and by this present testimony, I shall add the following names: Guillaume Duchoul,[13] a Prefect of Hautes Alpes in the region near Savoy and the Dauphiné; Bonifazio Rodigino, a Doctor of Jurisprudence and a distinguished astrologer; Giorgio Porro, a Rhætian; Luca Giustiniano of Genoa; Gabriello Aratore of Caravaggio, a noted arithmetician.

I entertained a warm affection for Giampietro Albuzzo, a physician and professor of Milan,[14] for Marc Antonio Morago, and likewise for Mario Gesso of Bologna. The fidelity of Lorenzo Zehener, a doctor of Carinthia, and of Adrian, a Belgian, toward me was extraordinary, and their services and kindnesses manifold.

Last of all, the patronage of the Prince of Matelica[15] was almost like divine protection, and greater than any which mere human devices would seem to be able to produce. I here barely touch upon the exceptional qualities of his soul, worthy, indeed, of a king; his knowledge of all subjects and of all teachings; the refinements of his spirit and his benignity. I but note in passing the vast accumulations of his wealth, the splendor of his father and his wisdom surpassing the peak of human attainment, and his memory for either favors or former associations.

What was there in me that was able to attract the gentle company of such an one? No benefits could I confer; nor was there any hope of something I might undertake for him, for I was an old man, bereft of fortune, despondent, and not especially gracious: It could have been nothing but his opinion of my character. Are you not inclined to deem such men godlike? They are those who devote as much energy to fostering love of study, to cultivating simplicity of living with a spirit gentle and true, to attempting and undertaking ventures worthy of praise, as almost every other man is wont to bestow upon the acquisition of power and favors and upon the nourishing of hope for the future, upon the daily round of custom and to the flattering of the great.

16.
CONCERNING MY ENEMIES AND RIVALS

I SHALL NOT undertake the same lengthy account of my enemies and rivals as I have just given of my friends. In this respect I think Galen erred not a little for he made the Thessalian, whose argument he took up, prominent by merely mentioning his name.

And if a man be not a coward, better it is to be reconciled, if you have been sinned against; better not to take revenge. Or rather it is better to take revenge by outdoing them than by resorting to invective. Therefore I have learned not so much to despise my rivals as to have pity on their vain-glorying. They bear witness that those men are far baser who conduct themselves underhandedly in their charges than those who act openly, if there is any just right to make accusation.[1]

17.

CALUMNY, DEFAMATIONS, AND TREACHERY OF MY UNJUST ACCUSERS

THERE ARE two kinds of treachery: one which spreads its snares around our reputation and honors, concerning which I have here planned to write; and a second, about which I shall treat later. Now, therefore, I have decided to discuss these perfidious designs, and especially secret designs; for no matter how treacherous such plots are, they have never been secret if the machinations are openly devised; if they are plots of any magnitude, they are, with difficulty, kept secret.

A man is a fool who attaches too much meaning to insignificant events; wherefore I shall content myself with the account of but four incidents.

The first of these befell me when I was summoned to Bologna.[1] The defamers of my name had sent a certain clerk to Pavia. This man, although he neither saw the inside of my lecture room, nor interviewed any of my students, managed somehow or other to write back to Bologna reports consisting of long tales which he had obtained from one who never thought the story would come out. The following is an extract: "Concerning this Girolamo Cardano I have learned that he lectures to the empty benches, since no students attend his classes; that he is a man of bad manners, disagreeable to all, and, in the main, a fool. He is given to disgraceful practices, and does not show even tolerable skill in medicine, being so given to certain prejudices in this art that no one engages his services, and he accordingly has no practice."

These statements were read by the Intelligencer at Bologna in the presence of the Most High Cardinal Borromeo,

Pontifical Legate of that city. Deliberations were afoot to pursue the matter no further, but when in the midst of the reading the committee had heard these words "he has no practice," one man present said, "Ho, I know that's a lie, for I have seen some very influential men who engaged his service, and although I myself am not a citizen of rank, I have consulted him."

The Papal Legate, Borromeo, thereupon, spoke up quickly, "I am abundantly able to testify," he said, "that Cardano cured my mother when her condition was given up as hopeless by all who attended her."

Then the former rejoined that perhaps there was only as much truth in the other statements as appeared in this, to which the Cardinal agreed. The Intelligencer was silenced and had the grace to look ashamed.

Consequently they came to this decision, that I should enter upon the office of professor for one year only, following the action of the committee; after that, quoting the terms, "if he will prove himself the sort of man described in the letter or otherwise unprofitable to the Academy and to the City of Bologna, let us suffer him to seek a post for himself elsewhere. But if we decide otherwise, the terms of the contract may later be confirmed, and the stipend established." There was already a disagreement over the salary. The Pontifical Legate assented, and so the business was concluded.

Not content with this, the committee directed a delegate from the Senate to meet me for the purpose of discussing the conditions of our agreement. He was to secure a modification of the terms already made, but I would not comply. He offered a smaller salary; no classroom for lecturing was definitely assigned; traveling expenses were not included. When he could get no satisfaction from me, he was forced to withdraw, and to return with all the original conditions stipulated in my contract.

Although such treachery may seem an impediment, and

calculated to thwart its object, this notion arises from a false idea of men, since all the ends of mortal activity can anticipate nothing more than a brief foreclosure, and are by no means as things everlasting. It is enough for the philosopher if he merely bears in mind the existence of such things and does not try to remedy them. The half of them amount to exactly nothing—not even the shadow of a dream, as anyone may observe and clearly see in his own actions; therefore they should not be taken seriously, since they have their limitations, and are precisely as significant as the games boys play with nuts for each player. If the boy thinks the outcome of these childish contests has something to do with winning laurels, or civil honors, or even a kingdom which will come to him when he grows to be a man because of some connection such as cause or precedent—how can he be other than a great fool?

After the preceding events, when I had at length entered upon my professorship, my opponents managed to dispossess me of my lecture room by this scheme: my lecture was scheduled for a period near the breakfast hour, and then the room allotted at that hour to another teacher. I offered three suggestions as against this vexation: that the other professor should begin and close his lecture earlier; or that he should give up the room entirely and permit me to be free to lecture in the room assigned; or else that he should use the room and I should be at liberty to choose another. When I saw that he was reluctant to take up with any of these propositions, I had it arranged at the next election that he should hold his classes elsewhere.

Out of this arose trouble and tears; one case of accusation after another interrupted my work, since I frustrated their conspiracies, and they were forced to see a man whom they hated lecturing.[2] Eventually, toward the end of my contract, rumors were circulated and even deliberately carried to the ears of Cardinal Morone, to the effect that only a handful attended my classes. This was not at all true. Indeed, from the opening of the session until Lent, I had had many auditors. Consequently, assailed by so many envious rivals and surrounded by innumerable plots, I thought the greatest virtue lay, as they say, in yielding to the force of circumstances.

Therefore, under the guise of consulting my honor, my enemies persuaded the Cardinal that I ought to resign voluntarily. They secured his action in the matter; and so the affair went through, more in compliance with the will of those who so urgently desired my removal, than in any wish for my happier lot.

Henceforth I am determined to say nothing on the subject of calumny and malicious slander; it is an evil so great, so persistent, and withal so contemptible and so irrational that it relies on nothing except far-fetched rumors and hinted accusations. And perhaps it was more hurtful for my calumniators to have been tormented by their own guilty conscience than for me to have been thwarted. For truly, they have left me more time for collecting my literary works; they have increased my fame; they have lengthened my life in releasing me from labors beyond my strength; they have provided me with an opportunity to enjoy myself, and devote myself to the investigation of many things not fully revealed to man.

I am, therefore, in the habit of saying—nay, the word is ever in my mouth—that I do not hate them nor regard them as culpable because they had their evil way with me, but because it was their evil purpose to thwart me.

What other bitter vicissitudes came down upon me, even before I was invited to Bologna, I shall set forth later in this account, in chapter thirty-three.

18.

THOSE THINGS IN WHICH I TAKE PLEASURE

AMONG THE things which please me greatly are stilettos, or *stili* for writing; for them I have spent more than twenty gold crowns, and much money besides for other sorts of pens. I daresay the writing materials which I have got at one time and another could not be bought for two hundred crowns. Besides these, I take great pleasure in gems, in metal bowls, in vessels of copper or silver, in painted glass globes and in rare books.[1]

I enjoy swimming a little and fishing very much. I was devoted to the art of angling as long as I remained at Pavia and I am sorry I ever changed.

The reading of history gives me extraordinary satisfaction, as well as readings in philosophy, in Aristotle and Plotinus, and the study of treatises on the revelations of mysteries, and especially treatises on medical questions.

In the Italian poets, Petrarch and Luigi Pulci,[2] I find great delight.

I prefer solitude to companions, since there are so few men who are trustworthy, and almost none truly learned. I do not say this because I demand scholarship in all men—although the sum total of men's learning is small enough; but I question whether we should allow anyone to waste our time. The wasting of time is an abomination.

19.

GAMBLING AND DICING

PERADVENTURE in no respect can I be deemed worthy of praise; for so surely as I was inordinately addicted to the chess-board and the dicing table, I know that I must rather be considered deserving of the severest censure. I gambled at both for many years, at chess more than forty years, at dice about twenty-five; and not only every year, but—I say it with shame—every day, and with the loss at once of thought, of substance, and of time.[1]

Nor was the smallest ground for defense left me. If, nevertheless, anyone may wish to rise in my defense, let him not say that I had any love for gambling, but rather that I loathed the necessities which goaded me to gambling—calumnies, injustices, poverty, the contemptuous behavior of certain men, the lack of organization in my affairs, the realization that I was despised by many, my own morbid nature, and finally the graceless idleness which sprang from all these. It is a proof of the foregoing assertion that once I was privileged to act a respectable part in life, I abandoned those low diversions. Accordingly it was not a love of gambling, not a taste for riotous living which lured me, but the odium of my estate and a desire to escape, which compelled me.

Although I have expounded many remarkable facts in a book on the combinations in chess, certain of these combinations escaped me because I was busy with other occupations. Eight or ten plays which I was never able to recapture, seemed to outwit human ingenuity, and appeared to be stalemates.

I have added these remarks to advise any who may by chance happen upon the same extraordinary situations—and I hope somebody may—so that they may add their jot or tittle to the solution.

20.
DRESS

OF MYSELF I have the same opinion as Horace had of his Tigellius; rather I might have said Horace was then speaking of me in the character of Tigellius:

> Nil æquale homini fuit illi; sæpe velut qui
> Currebat fugiens hostem; persæpe velut qui
> Junonis sacra ferret; habebat sæpe ducentos,
> Sæpe decem servos; modo reges, atque tetrarchas,
> Omnia magna loquens; modo, Sit mihi mensa
>   tripes et
> Concha salis puri, et toga quae defendere frigus
> Quamvis crassa queat.[1]

You ask why this comparison; the reasons, forsooth, are ready. First, the variety of my interests and my habits coupled with the fact that I am ever solicitous for the health of my body. And since I have often moved from one country to another, I have found it necessary to make a change in the arrangement of my garments. This would leave on hand clothes which it did not seem desirable to sell on account of the loss involved, or to keep to no purpose. Accordingly, necessity dictated my practice in dressing.

A second consideration no less important than the latter, though less urgent, is my own indifference towards domestic or personal matters in the interests of study, and the consequent neglect of the care of my clothes because of the remissness of my servants. In this way, a large wardrobe is apt to be reduced by wear to a few rags before I notice it.

Wherefore, I by no means disdain the advice of Galen, who taught that a man ought to be content with four sets of garments, or only two, if you do not count his undershirts. However, I am of the opinion that such garments, after they have done their duty for a given occasion, may be interchanged, and indeed ought to be.

I think four suits would be sufficient for any man of us, so that we could wear our choice of two heavy garments, one medium in weight, and the other very warm; and of two lighter, one, likewise, medium in weight, and one very thin. With these, fourteen combinations of apparel would be possible, not counting the arrangement by which all could be worn at once!

21.

MY MANNER OF WALKING AND OF THINKING

BECAUSE I think as I walk, my gait is uneven, unless somebody claims my attention. My feet are moved, and often my hands, even, make gestures at the bidding of my restless mind. The very diversity of my concerns, the circumstances that befall me, nay, the very disposition of my body, influence my going. When I am well, when I am buoyant and not weary, I hasten along carefree and happy; contrary circumstances retard my steps. My gait, therefore, is likely to resolve itself into a subject for comment—become a by-word —for I stagger along heedlessly, meditating on many a topic alien to my surroundings.

In general, all things which a hard necessity controls are variable; yet an impulse of the mind governs each man, so that he can persist in what is good and be unwilling to hold to evil. This can be accomplished only by constant application of thought, though not necessarily always upon the same themes. Nevertheless, sustained thought takes such complete possession of me that I may neither eat, nor enjoy myself, nor even succumb to grief or sleep without attendant meditation. This, then, is a great good which may ward off evil and offer relaxation, and yet should it cease, I know not whether the result would be a help or a hindrance.

For the rest, my walk is not quick, nor slow; now I proceed with head and shoulders erect, and again I go bent, differing little, and in this respect especially, from the manner of my gait in my youth.

22.

RELIGION AND PIETY

THOUGH I was born in most troublesome times,[1] and have been subjected to the influence of so many experiences; though in my many journeys I have met men not simply strangers to religion, but indeed the enemies of religion, I have not lost my faith; and this I must attribute more to a miracle than to my own wisdom; more to Divine Providence than to my own virtue. Steadfastly, in fact from my earliest childhood, I have made this my prayer: "Lord God, in thine infinite goodness grant me long life, and wisdom, and health of mind and body." Wherefore it is no marvel if I have ever been most devoted to religion and to the worship of God.

Indeed, it seems that I have been the recipient of other gifts as well—such things, however, as are apparently more another's needs than my own. At all times, I have enjoyed a certain health, in spite of my complaints. I have become more learned, so to speak, in matters to which I gave little study, and in which I had no schooling, than in subjects for which I had resorted to teachers. Where duty was concerned, I was somewhat more assiduous; I fought against my son's death and that terrible grief; yet he was bound to die, and, in that same year, the infant he had left came near to dying, so that my son would have had no issue, but my grandson—his child—lived and still lives.

Yet why moan on? Why compare mortal wretchedness and pain to the joys of those who have won immortality? If he had never existed at all! Or if he had not perished at that time, would he have lived forever? What difference then, to me, if I have suffered some loss? Oh, senseless soliloquies of men! Oh, unutterable delirium!

Not only do I keep ever in mind the divine majesty of God, but turn my meditation as well upon the Blessed Virgin Mary, and holy Saint Martin; for I had been warned in a dream that under his ægis I should lead a quict cxistence and enjoy long life.

At one time I wrote a long essay which I here give in summary, on the theme that the evils of this life can in no manner be equal to that bliss which we trust is in store for us; that we find no difficulty at those times when unusual circumstances of a certain sort surround us, in being so moved that doubt is not possible, and we feel that everything is as it should be. But when these circumstances fall away, everything seems to be a dream.

Oh, would that, had it so pleased God, this Charybdis of uncertainty might have given place to a sense of duty as great. For if men would have reverence for the commands of heaven, if they considered how richly they might enjoy the blessed satisfaction of having remembered to observe God's will, they would live more devoutly and would lead exemplary lives. But I realize that I am laboring at an extremely unpopular task in wishing to ordain for mortals a commandment of wisdom. In this my devoutness sweeps me on, and sorrow for men of wretched estate. Accordingly, I consider that, in the number of those who have made their humble contributions to a discussion of the immortality of the soul, I have written quite naturally, and in a manner by no means at variance with Plato, Aristotle, or Plotinus, and in no way opposed to reason or understanding.

Of Plato, moreover, a certain gravity is characteristic; Aristotle is gifted with logical or rhetorical division; in Plotinus one feels the lack of definition and applications. Of this last, however, I was not the discoverer, but cheerfully ascribe the criticism originally to Avicenna[2] who is much like him among the philosophers.

23.
MY OWN PARTICULAR RULES OF CONDUCT

IN NO RESPECT have I shown myself more adept than in formulating my experiences, partly because of my long life, partly because of my numberless adversities. First, then, aside from those inept little prayers of my childhood when I was just beginning to learn, I have been in the habit of returning thanks to God for all that has happened to me. I have thanked him for the gracious favors he has bestowed; indeed I should consider anyone base and a brute who would not return gratitude even to man for such blessings. I have thanked him in adversity, and have accepted the light afflictions as admonitions wherefrom I might take warning. Oh, how many times, taught by my trials, have I avoided the most devastating calamities!

For the barely tolerable troubles of my life—even to these I feel I am indebted—I return him thanks, because I believe that none of these things which time is wont to obliterate can be momentous; and I recognize that God is the dispenser of all my afflictions which, though they may have seemed oppressive at the time, I doubt not were very good in the great order of the universe. Granted that death is inevitable, a host of disasters may render it easier to bear. As Ægineta[1] used to say: "He who passes a large stone from the bladder suffers less, by contrast with his preceding pains, than he who passes a small gravel, and he is less likely, therefore, to perish." Even in the very extremity of my suffering I am persuaded that God did not forget me, and with this assurance I have fought off death—*mirabile dictu!*—in the midst of death.

My second noteworthy observation was that I ought ever to pray to know the purpose of God; even in accordance with the Scripture I invoke the spirit of high God, that he may teach me to do his will, because he is my God. Behold how good, how pleasant are his ways! I was sustained by threefold comfort in my calamity: for he gave before he took away; he kept me safe against the sweeping billows of life's sea; he granted me a tranquil existence.

A third rule was that when I had lost, I should not be content merely to redeem the loss, but should always obtain something in addition. I am, no less, one of few men who enjoy experimenting with life as well as acting from carefully deliberated motives.

In the fourth place, I made it a practice that I should take the most careful account of my time. As I rode or ate or conversed, or as I lay in bed sleepless, I was ever meditating upon something, for I had in mind that common adage: "Multa modica faciunt unum satis." That is, the many small things soon make one of size!

Here I shall tell a brief tale, and a true one. While I was living in the Ranuzzi mansions in Bologna, there were two apartments, one of which was dingy but safe; the other was gorgeously decorated as to the walls, but over them hung a dilapidated ceiling, threatening momentarily to collapse. Parts of it had fallen while I was living there, with no little danger to my life, had I been underneath; and when finally a shower of fragments came down at once, I barely escaped, with my head intact, to safety.[2]

By a fifth rule I observed that it was well to cultivate the society of elderly men and be with them frequently.

A sixth practice was to observe all things, and not to think that anything happened fortuitously in nature; whereby it has come about that I am richer in the knowledge of Nature's secrets than I am in money.

Always to set certainties before uncertainties has been a

seventh guiding principle; and as a result of this I have been so fortunate that I am persuaded I owe most of the events of lucky issue in my career to this resolution.

My eighth axiom bids me never to be persuaded, for any reason whatever, to persist willingly in any course which is turning out for the worse; in this I pay tribute to experience rather than to my own wisdom or to any overconfidence in my own skill; this is especially true in the cure of the sick; in other things I have been willing to intrust myself to chance. I do not regret what has once been done, saying as do many, "What if I had done thus and so?" What is the good of this? To question too much wherein lies the greater gain may prove no gain at all because of expenditure of time.

In caring for the sick you should never regard action as second to efficacy: that is, you should prefer an irrigation to an untreated fistula; you should not neglect the water of the intestines in a case of dropsy. When, after employing a rather more potent remedy, or one of the customary treatments, especially if it be one held in good repute, your results are not what you expected, you should then proceed so much the more gently; for, as I have always maintained, all remedies should be suitable.[3]

Unless I am at leisure, I undertake no particularly disputatious engagements, not only because it is more expedient, but because in this way I do not waste time.

I never slash at a friendship that has proved faithless; I gently ravel the threads that have woven our interests together.

Shortly after I was seventy-five I expressed my unwillingness to be present for a fee at a consultation unless I knew how many or who my associates were to be.

Flee any occasion to let familiarity breed contempt.

As far as I have been able, I have trusted less to my memory, and more to the written word.

24.
MY DWELLING PLACES

WHEN I was a child my home was at Milan in the Via dell' Arena by the Pavian Gate.[1] From there we moved to the Via dei Maini[2] which was in that part of the city near the castle, and later we rented a house of Lazzaro Soncino. That was when I was a small boy.

During my youth we lived in the Via dei Rovelli in a house belonging to Girolamo Ermenolfo, and following that, in a property of the Cusani. In young manhood, up to my nineteenth year I lived in Alessandro Cardano's house.

At Pavia I occupied successively, a house near the church of San Giovanni in Burgo,[3] one near the church of Santa Maria di Vénere[4] which belonged to the Catanei, and another near the church of San Gregorio in Monfalcono in Burgoliate. Later I moved to a place near the University halls, adjoining the residence of the deputy Ceranova, and shortly after bought a house of my own near the church of Santa Maria in Pertica.[5]

At Bologna I first took a place in the Via de Gombru,[6] but left it for the Palazzo Ranuzzi in the Via del Galera. Later I purchased a property near the church of San Giovanni in Monte.[7]

At Rome I first had residence near the Porta del Populo in the square of San Girolamo hard by the Curia Savelli. My next home in that city was in the Via Giulia near Santa Maria di Montserrato.

Long before this I had once lived in a house my mother

had bought in Milan adjoining the church of San Michele di Chiusa. Thence we went to the Porta Orientale; from there to the Via Cinque; and finally, from a rebuilt house which had once fallen down, I returned to the first, near San Michele.

25.

POVERTY AND LOSSES IN MY PATRIMONY

I WAS POOR, yet not greedy for gain; nor did I strive after the vain and ostentatious grandeur of outward show. My domestic concerns suffered a fall because of war in my country accompanied by excessive tax levies, and because I was almost never without a numerous household. I was calumniated by my bitter rivals; for a long time the College of Physicians at Milan refused to recognize me. I was more than once a heedless wastrel. My body was a weak thing, and my affairs tottered upon foundations undermined by frauds. I spent money lavishly for books, and wasted much substance in moving so often, whether from city to city, or from one residence to another.

The time I trifled away at Gallarate was profitless; in nineteen months I scarcely earned twenty-five crowns toward the rent of my house. In a turn of ill-luck at dicing, I put to pawn my wife's jewelry and some of the furniture. And though it is confounding to admit that I was capable of squandering our very substance, it is more surprising that in my destitution I did not take to begging; stranger still, that I never even seriously considered that my course was an insult to my ancestors, to a decent standard of manhood, and to the honors which I had achieved, and by virtue of which I was later to become prosperous. I went my reckless way serenely.

This state of affairs left its imprint upon my career for the next fifteen years, nor in this time did I exhibit any desire to profit by the offices of a practicing physician.

"Indeed," you will say, "how did you support yourself? Did you give private instruction?" "No, I did not."

"Did you accept a loan without security?" "No."

"Did you not apply to someone for a gratuity?" "No. I doubted that I should get it, and I was ashamed."

"Perhaps you cut down on your living?" "Not that either."

"What then?"

"I wrote almanacs; I lectured under the Plat Endowment in the public halls of instruction. I got together a little by practice of medicine; my several house-servants were resourceful. The family of Archinto patronized me at times with small donations, and I sold prescriptions. I kept my eyes open for all contingencies, and imitated the gleaners of the field. Upon fine clothes I steadfastly turned my back.

"Thus I bore the fortune of those difficult days and prepared myself the better to play my part in the enjoyment of prosperity."

26.
MARRIAGE AND CHILDREN

MANY YEARS ago I dwelt in the town of Sacco, a gay and happy young man, free, as freedom goes, from all cares, a mortal in the seats of the high gods as it were, or better said, in the realms of bliss. At the risk of being irrelevant I must at this point refer to a dream which is too appropriate to my theme to pass over. On a certain night I seemed to be in a delightful garden of surpassing charm, gay with flowers and abounding in fruits of divers sorts; a gentle breeze was murmuring. No painter could have depicted anything more lovely, nor could Pulci, the poet, have captured it in song, nor the imagination have framed its like.

At the entrance, moreover, of the garden, a gate stood open; a door facing it also opened, and lo, in my presence stood a girl dressed in shining white. I seemed to embrace her and kiss her, but at the first caress, a gardener immediately appeared and closed the door. Earnestly I began to plead with him to leave it open, but he would not. Sadly then, and clinging to the girl, I seemed to stand there, shut out.

A few days after I had dreamed this, a house in the town burned, and I was roused up at night to go to the fire. I knew who owned the house—one Aldobello Bandarini,[1] an officer of the Venetian Militia levied in the country around Padua. I was not greatly concerned. I scarcely knew the man by sight, but I was not at all pleased when, by a mere chance, he rented the house adjoining my own; he was not the sort of neighbor I would have desired, but what could I do?

Meanwhile, after an interval of some days I saw, from the

street, a maiden who exactly resembled in face and fashion of dress the girl of my dream. "Oh," said I, "what have I to do with this maiden? If I, a pauper, marry a wife who has no *dot* save a troop of dependent brothers and sisters, I'm done for! I can scarcely pay my expenses as it is! If I should attempt an abduction, or try to seduce her,[2] there would be plenty to spy upon me. Her father, a fellow-townsman and an officer of militia, will never tolerate any violence of that sort, and, in any case, what course would I adopt to carry out such a design? Alas, for the affair to turn out well at all, I should be obliged to flee."

These considerations, and others like them, filled my mind as I examined every possibility; I felt it was better to die than to go on living thus. From that day, I ceased merely to love the girl; I was fairly consumed with passion. I knew how much I dared hope for from the reading of my dream; I knew I was free from the bonds of impotency.[3] Nothing loath I married her, for she was willing too; her parents were even soliciting the match, and making offers of help, should any be needed, for help could do much for us just then.[4]

Truth to tell, the interpretation of my dream did not find its conclusion in the person of this girl, but revealed the full force of its meaning in my sons. For this unfortunate union was the cause of all the calamities which befell me throughout my whole life. Whether these ills were a dispensation of Divine Providence, or whether they were due me through my own or my forebears' sins I know not. I only know that otherwise I have steeled myself to rise above any misfortunes which may overtake me.

27.
THE DISASTERS OF MY SONS

IT WAS IN my sons that the full force of my dream was manifestly revealed. First, after having twice miscarried and borne two males of four months, so that I began to despair of issue, and at times suspected some malefic influence at work, my wife brought forth my first born son—a child exactly resembling my own father in features. In his youth he showed himself good, kind, and simple-hearted. He was deaf in his right ear and had small, white, restless eyes. Two toes on his left foot, the third and fourth counting from the great toe, unless I am mistaken, were joined by one membrane. His back was slightly hunched, but not to the extent of a deformity. The boy led a tranquil existence up to his twenty-third year. After that he fell in love, about the time he was granted his degree, and married a dowerless wife, Brandonia di Seroni.

His mother, as I have said, had long since passed away; and much before that, even, his maternal grandfather, who, as a matter of fact, had survived my marriage to his daughter but a few months. Only his mother's mother, Taddea, was still alive.

Then in truth came grief and tears! For in times past, while his mother lived, I had patiently suffered much, but these ills had finally had an end. I refer to the time when I was enduring the attacks of my enemies. My son, between the day of his marriage and the day of his doom, had been accused of attempting to poison his wife while she was still in the weakness attendant upon childbirth. On the 17th day of

February he was apprehended, and fifty-three days* after, on April 13th, he was beheaded in prison.

And this was my supreme, my crowning misfortune; because of this, it was neither becoming for me to be retained in my office, nor could I justly be dismissed. I could neither continue to live in my native city with any peace, nor in security move elsewhere. I walked abroad an object of scorn; I conversed with my fellows abjectly, as one despised, and, as one of unwelcome presence, avoided my friends.

What course to take failed to present itself; I had no place to withdraw; I know not whether I was more unhappy than hated.

Hard upon this succeeded the folly, the ignominious conduct, and violent actions of my younger son until nothing could have contributed to his further disgrace. I was obliged to have him imprisoned more than once, to condemn him to exile, and to cut him off from his paternal inheritance; nor was there any property from his mother's side.

From my daughter alone have I suffered no vexations beyond the getting together of her dowry; but this obligation to her I discharged, as was right, with pleasure.

From my eldest son I took two grandchildren to rear, yet one house—mine—witnessed, within the space of a few days, three funerals, that of my son, of my little grand-daughter, Diaregina, and of the baby's nurse; nor was the infant grandson far from dying. In general, everything pertaining to my children had the worst possible issue. Even my daughter, in whom at least some hope of good had resided, married as she was to the prosperous and distinguished young Bartolomeo Sacco, a citizen of Milan, was barren; thus the sole hope of succession rested in my grandson.

I am by no means unaware that these afflictions may seem

*Cardano again makes an error in computation; there are fifty-five days between the dates.

meaningless to future generations, and more especially to strangers; but there is nothing, as I have said, in this mortal life except inanity, emptiness, and dream-shadows. What is that basis on which, more than any other, all acts of mortals, all their affairs, their very life and their vicissitudes may find a firm foundation? But, as in Cranto,[1] Cicero, the father of eloquence, shows how he may find consolation in the death of his daughter, so, out of the greatest adversities of this sort, mortal things may find, now here, now there, new meaning and testify that they are destined for a purpose and a use not to be despised.

For the rest, it is clear to me that only those things seem noteworthy which are generally recorded in books, such as a series of great events arising from the most insignificant beginnings. In recounting these the narration should move briskly and each incident be set forth with the most scrupulous order, so that the finished picture may be a fair representation of actual history. Or again, the portrayal of conspicuous performances, which have come about through one man's lofty character or another's baseness, or by mere chance, should be told as succinctly as possible, unless a discussion of art, or philosophy is involved.

But today—O tempora! O mores!—we write naught save servile flatteries. It is legitimate to extol those worthy of it on the score of high character and blameless life, as Pliny honored Trajan and Horace honored Mæcenas. But we, stupidly, go about explaining how the thing should be done. If only a man could realize that this flattery is in no way praiseworthy, but rather most disgusting, like two mules who scratch each other. If, however, it is really deserved, then give it a word in passing, as if it were a well known fact, just as Pliny the Younger did when he referred to Martial.

A book which is worth buying ought to seek perfection in wisdom as well as in art. That is a perfect book which, pursuing a single theme from the beginning, arrives thereby at a logical conclusion; it neither omits any relevant point, nor includes the irrelevant; it conforms to the rules of rhetorical division; it offers some explanation of recondite matters and gives evidence of the fundamental principles on which its argument is based; or it may be a work which accurately interprets a master of some art, as Vitruvius has been interpreted by Philandrier.[2]

28.

PROCESSES AT LAW

FROM THE death of my father up to my forty-sixth year, that is, for twenty-three years, I was almost constantly engaged in suits at law. The first, with Alessandro Castiglione, called Gatico, was in connection with certain woodlands. This suit was afterwards continued with his kinsmen. The next litigation was with the Counts Barbiani; following that, with the College of Physicians at Milan, and finally with the estate and heirs of Dominico de Torris, who had held me at the baptismal fount. I won all these cases. It was a marvel that I should obtain a verdict against Alessandro Castiglione; his uncle was the presiding judge, and he himself had obtained a judgment against me which had been reversed in the process, in the language of jurisconsults, and I forced him to pay all money due me.

By a similar turn of fortune I was first considered as a candidate by the rectors of the college at Milan, and excluded by a majority vote; my admission was then ratified at the final election, and I was made an associate, subject to the college. I was not, however, admitted on equal terms though eventually received full privileges in the face of a strong opposition.

I came to an agreement with the Barbiani also, after long process, threats and other obstacles, and, having received the stipulated sum, was at last entirely free from the law-court.

29.
JOURNEYS

IN THE COURSE of my many journeys I have visited almost the whole of Italy, excepting Naples, Apulia, and the regions neighboring these. Likewise I have seen Germany, especially lower Germany, Switzerland, and Tyrol; besides these countries, I have been in France, England, and Scotland; and I will tell how this happened.

John Hamilton,[1] Archbishop of St. Andrews, was a chief of state in Scotland. He was natural brother to the regent, Pontifical legate and primate. He suffered with periodic attacks of asthma, recurring, in the beginning, at somewhat —lengthy intervals. When he had reached his fortieth year, the interval was reduced to eight days, and death seemed imminent, inasmuch as in twenty-four hours, he was eased by no, or at best by very slight, relief. He repeatedly consulted, but in vain, the court physicians of Charles V, then Emperor, and of Henry, King of France. Finally, my name having come to his ears, he sent, through the intermediation of his own physician, two hundred crowns to me in Milan, to the end that I should proceed to Lyons, or to Paris at most, implying that he would come there. I, since I was not at that time engaged in lecturing, as I have explained above, willingly welcomed his terms.

And so in the year 1552 on February 23, I was ready to set out upon my journey, crossing by way of Domo d' Ossola, Sion, Geneva, and Mount Simplon; and, leaving Lake Geneva behind, I arrived in Lyons on the 13th of March. It was during the Milanese carnival, on the sixth day, by common reckoning.

There I tarried forty-six days, but not a glimpse of the Archbishop, nor the physician himself, whom I was awaiting. Meanwhile, my fees for services much more than paid my cxpenses. Lodovico Birague, the distinguished Milanese citizen, with whom I had long maintained an intimate friendship, was there, serving at that time as captain of the royal infantry. He went so far as to come to me with an offer of a yearly stipend of a thousand crowns if I would enter the service of Marshal Brissac.[2] During this time, William Casanate, physician to the Archbishop, arrived, bringing with him an additional three hundred crowns which he gave me in order that I might be encouraged to proceed to Scotland. He offered to defray all my traveling expenses thither, and made promises of additional liberal rewards.

Accordingly I was conducted across the country, by way of the River Loire, to Paris. There I happened to see the great Orontius,[3] but he refused to visit me.

Nicholas LeGrand[4] took me to see the private treasure vaults of the King of France at the church of St. Denis, a place of no very great fame, but rather the more noteworthy in my estimation especially because of the perfect horn of a unicorn preserved therein.

Thereafter, a conference was held with the physicians to the King. We dined together, but they did not succeed in getting an expression of my views at table, because before the meal they had wished me to take precedence in expressing my opinion.

I continued my journey on the best of terms with Jean Fernel, Jaques de la Boë,[5] and another court physician, all of whom I left regretfully. I went on to Boulogne in France; whence, escorted by fourteen armed riders, and twenty soldiers, at the order of the Prince of Sarepont, I traveled on to Calais. There I saw the tower of Cæsar still standing.

Therefrom, having crossed an arm of the sea, I went to London. At length I came to Edinburgh, to the side of the

Archbishop, on the 29th of June. I remained there until the 13th of September. For my services I received another 400 gold crowns, one neck-chain worth 25 crowns, a thoroughbred riding-horse, and many other gifts, so that not a single member of my party went away empty-handed.

Returning, I came first to Brabant, and in the region about Tongres I visited Gravelines, Antwerp, Bruges, Ghent, Brussels, Louvain, Malines, Liège, Aix-la-Chapelle, Cologne, Coblenz, Kleve, Andernach, Mainz, Worms, Spires, Strasburg, Basle, Neustadt, Berne, Besançon, crossing through the heart of the Tyrolese state, and visiting Chur and Chiavenna, cities of this region. Finally, crossing Lake Como, I reached Milan on January 3, 1553.

Of all these places, I tarried only in Antwerp, Basle, and Besançon. My friends in Antwerp made every effort to retain me.

At London I was granted an audience with the King, and while there accepted one gift of an hundred crowns, and rejected another of five hundred—some say a thousand, the truth of which I am not able to ascertain—because I was not willing to give acknowledgment to a title of the king in prejudice to the Pope.[6] While in Scotland, I became an intimate friend of the Duc du Cell, Viceroy of the French.

At Basle I almost took lodging in a hostelry infested with the plague; and, indeed, would have done so, had I not been warned by Guglielmo Gratarolo. At Besançon I was hospitably received by the Prelate of Lisieux, as I have already noted; he presented me with gifts, as was the case elsewhere.

Altogether, I have lived four years at Rome. I spent nine at Bologna, three at Padua, twelve at Pavia, four at Moirago—the first four of my life—and one at Gallarate. I was in the town of Sacco almost six years, and in Milan thirty-two. I made three successive moves within three years.

Besides my long journey to Scotland, I went to Venice and Genoa, and visited the cities which are on the way—

Bergamo, Crema, Brescia, and others. I went, as well, to Ferrara and to Florence, and beyond Voghera and Tortona.

Briefly, I may say I am familiar with nearly the whole of Italy excepting the Kingdom of Naples and states adjoining, as the old Apulian region, ancient Latium, the Marches, Umbria, Calabria, ancient Magna Græcia and Lucania, and Abruzzi.

But, you will ask, to what purpose is this account of all these cities? There is great value: for if you will look about you only a single day, according to the suggestion of Hippocrates, you will know what may be the nature of the place and the customs of the inhabitants, what section of the city it is better to choose, and what diseases are prevalent. We may also determine which of the regions we visit is more favorable, for at one time an entire district will prove scarcely habitable because of the cold, and again another district undesirable on account of troublesome times.

An acquaintance with other lands is, besides, profitable for the better understanding of history, and especially helpful to mathematicians writing treatises on geography, or to anyone interested in the nature and productive usefulness of plants and animals. Again it gives one a knowledge of the routes mainly traveled; and out of the experiences of travel many books on this subject are published in Italian, thus presenting additional information about things far distant.

30.

PERILS, ACCIDENTS, AND MANIFOLD, DIVERSE, AND PERSISTENT TREACHERIES

THE ACCIDENTS which I am about to relate happened to me when I was residing in the house of the Catanei at Pavia. One morning I was going to the University halls. Snow lay upon the ground. I had paused to relieve myself beside a ruined wall on the right of the school premises; and therefrom, as I continued my way through the lower part of the passage, a loosened tile tumbled perpendicularly. I avoided this peril only because I could not walk through the snow which filled the upper pathway, next the wall, where my companion had urged me to go.

In the following year, 1540, if I am not mistaken, when I was passing through the Via Oriental, it suddenly occurred to me, for no reason whatsoever, to cross from the left side to the right; and when I was once over, a great mass of cement fell from a very high cornice on the opposite side, precisely such a distance ahead of me that certainly, unless I had changed my course, it would have ground me to bits; thanks to God, I escaped.

Shortly after, near the same place, I, riding a mule, passed near a large wagon. I wanted to proceed toward the right, for my business was urgent, and the delay was annoying. I said to myself, "What if this wain overturns!" And before long, even as I was stopping there, it did fall, and there is no doubt that it almost crushed me. I suffered visible harm, and had run no small risk.

It is not on account of this single incident that I marvel at the outcome; but because on so many occasions I have changed the direction of my going, always involuntarily, except in perils of this sort, or perhaps in other dangers I have not been aware of. Nevertheless, it is not the significant event which ought to be wondered at, but rather the frequent recurrence of similar instances.

When I was a boy of eleven, if I am not mistaken, I was entering the courtyard of Donato Carchani, a citizen of rank, when a small furry dog bit me in the abdomen. I was injured in five places, not seriously; but the wounds were blackish spots, and in this accident I can say nothing less than that I was exposed to I know not what dangers of rabies. If this had happened to an older person, what the bite had not done, fright would surely have finished completely.

In 1525, the year in which I became rector, I was almost drowned in Lake Garda. Rather reluctantly I had boarded a craft which was transporting some horses for hiring. During the crossing, the main mast, the rudder, and one of the two oars with which the boat was manned, were broken. The sail was rent, and even a smaller mast snapped off; and at length night overtook us. I reached Sirmione in safety after the other passengers had abandoned even their faintest hopes, and I was all but desperate.

Had our embarkation been delayed but the fortieth part of an hour, we should have perished; for such a violent storm gathered that the iron bars of the window-shutters on the inn were bent. Though I had seemed thoroughly uneasy from the first, I contentedly ate a hearty supper once an excellent pike appeared on the table. The rest of the passengers were not so calm, excepting the counsellor of our disastrous voyage, who had, however, done efficient service in the moment of our peril.

Once when I was in Venice on the birthday of the Blessed Virgin, I lost some money while gambling; on the following day I lost the rest, for I was in the house of a professional cheat. When I observed that the cards were marked, I impetuously slashed his face with my poniard, though not deeply. There were in the room two youths, the body-servants of my adversary; two lances were fastened to the beamed ceiling; the key was turned in the door.

When, however, I had begun to win and had recovered all the money, his, as well as my own, and the clothes and rings which I had lost on a previous day, but from the beginning of the next day had won back from the start; and since I had earlier despatched these belongings to my lodging with my servant, I tossed a part of the money back, willing to make amends when I saw I had wounded him. Then I attacked the house-servants, but since they were unable to handle weapons, and were beseeching me to spare their lives, I let them off on the condition that they should throw open the door of the house.

The master, seeing such a commotion and tumult in his household, and anxiously fearing every moment's delay, I judge, because he had defrauded me in his own house with his marked cards, after making a rapid calculation of the slight difference between what he had to gain or what to lose, ordered the door to be opened; thus I escaped.

On that same day about eight o'clock in the evening, while I was doing my best to escape from the clutches of the police because I had offered violence to a Senator, and keeping meanwhile my weapons beneath my cloak, I suddenly slipped, deceived in the dark, and fell into a canal. I kept my presence of mind even as I plunged, threw out my right arm, and, grasping the gunwale of a passing boat, was rescued by the passengers. When I scrambled aboard the skiff, I discovered in it, to my surprise, the Senator with whom I had just gambled. He had the wounds on his face bound up with a dressing; yet he willingly enough brought me out a suit of garments such as sailors wear. Dressed in these clothes, I traveled with him as far as Padua.

Once in Antwerp when I went to a shop to purchase some

gems, I stumbled into a pit, which was in the shop for some reason or purpose I did not understand, and hurt myself. My left ear was scratched, but I made light of the accident, for I suffered but an abrasion of the skin.[1]

In the year 1566 while in Bologna, I was thrown from a careening carriage which could not be checked. In this accident the ring-finger of my right hand was broken, and my arm so badly injured that I was not able to bend it backwards. The stiffness remained for some days, and then passed over to the left arm, while the right remained unimpaired—a rather remarkable circumstance. But what is still more surprising is that nine years later, this stiffness, for no manifest reason, but rather like an omen, returned again to the right side where it troubles me at present. The finger, however, although I gave it no special treatment, had so healed that I suffer no inconvenience from it and am no longer annoyed by the crook in it.

What should I say about the risk I ran of plague in 1541? I had been summoned to attend the servant of a colonel, a Genovese gentleman of rank from the Island. The servant was in the grip of the plague, for he had just come from Switzerland where he had slept between two men, both plague-stricken, who had later died of it. Unaware of such serious circumstances, I bore, in my official capacity as Rector of the College of Physicians, the canopy of the Emperor on the occasion of his entry into Milan. When it was known that my patient at this very time had the plague, the colonel wanted to hide the evidence by hurriedly concealing the dead man—for he was considered as good as dead—in a place without the city. I was unwilling to give my assent to this concealment, fearing nothing more than to be the victim of a deception. But with the help of God the sick man recovered, contrary to all hope, and aided not a little by my almost fatherly care.

I scarcely know how to relate the most amazing occurrence which befell me in the year 1546. I was coming away, all care-free, from a house in which, on the previous day, a dog had snapped at me. He had not, however, actually bitten me to wound; but because he had sneaked upon me so slyly and quietly, I feared that he might have been a mad dog. Although he did not drink from a dish of water I set down for him, still, he did not flee from it, and he had eaten a chicken thigh which I had directed to be given him.

Well then, to continue: as I was coming along, I saw a very large dog approaching me, but still at a distance. It was on the Feast of Santa Cruz, in the month of April, and I was riding through an exceedingly pleasant way, closed in on either side by greening hedges and trees. I said to myself, "What affair have I with dogs? A dog yesterday; a dog today! I have never entertained silly fears, but who knows whether this dog is really mad?"

While I was considering these things the dog was loping along, straight for my mule's head, so that I hardly knew what to do next; as soon as he was close enough, he sprang forward, ready to attack me. I was mounted on a small mule, but immediately the idea of making the only move that could help matters occurred to me: I bent my head low upon the mule's neck, and the dog passed over my head, gnashing his teeth! He not only did not wound me, but did not even touch me—a fact which surely can be numbered among the miracles!

If I had not committed this story to writing by frequent reference to it in many of my books, I should think I had dreamed it. Or I should even suppose that I had suffered an hallucination brought on by my meditation, had I not by chance observed, in looking behind me, that the dog was running back and leaping at a boy who was following me on the left near the hedge. I returned and questioned the boy. "Tell me, I beg you"—for the dog had now gone on in headlong course—"whether you saw what that dog did? And has he injured you?"

"He did not hurt me in the least," replied the boy, "but I certainly saw what the brute did to you."

"Tell me, I beg you, what he did," I requested again.

"He sprang directly toward your head, and when you bent your head, cleared you without hurting you."

Then I said to myself, "For a truth, I am not wandering in my mind."

Yet this incident still seems beyond credence to all.

In general, then, I had four extremely perilous encounters. In a word, such hazards, that if I had not in some way circumvented them, it would have been a question of my life. The first was the danger of drowning; the second, mad dogs; the third and lesser because it was only an incipient danger, the falling mass of masonry; and last, the quarrel in the house of the Venetian noble.

I was the victim of just as many great discouragements and obstacles in my life. The first was my marriage; the second, the bitter death of my son; the third, imprisonment; the fourth, the base character of my youngest son. Thus systematically ought one's life to be examined.

I pass over the sterility of my daughter; the long strife with the College of Physicians; so many iniquitous, not to say actually unjust, implications brought against me; such vile uncertainty of bodily health and constant weakness; the fact that I had no associate whose friendship brought me good counsel and advantage. Had such a companionship not been lacking in my life, I might have been relieved in a large measure, and been saved the sting of many difficulties.

Attend now, while I relate certain episodes which, *mirabile dictu*, are concerned with a deadly plot; for it is, in truth, a curious story.

In the course of my teaching at Pavia, I was used to reading at home. At that time I had in the house a woman who rendered irregular service, the youth Ercole Visconte, two young boys, and a man servant, I think. One of the boys was an amanuensis and a musician; the other, a foot-page.

The year 1562 brought me to the place where I had determined to leave Pavia, and resign my professorship. The Senate viewed this decision with manifest disfavor, as if the case involved a man acting in anger. Now it happened that there were two certain doctors in Pavia. One was my former pupil, a crafty man; the other, a professor extraordinary of medicine. He was a guileless man, and not such a one as I deem mischievous; yet, forsooth, what does greed for position and power not accomplish, especially when it is allied to a very genuine zeal for achievement? There was also, moreover, that illegitimate opponent. These rivals were striving their utmost to force me to quit the city, ready to adopt any measures for the sake of fulfilling their avowed purpose. Accordingly, when they despaired of getting rid of me because of the Senate's objection, as against my own eagerness to withdraw, these adversaries plotted to murder me. This they arranged to do, not with a sword, for fear of scandal and the Senate, but by a deadly scheme. My competitor saw that it was impossible for him to be head unless I should withdraw. Therefore they went about their business indirectly.

First they composed a letter in the name of my son-in-law—a most infamous and vile letter to which they did not neglect to add, as well, my own daughter's name—declaring that he was ashamed of acknowledging our relationship; that, in behalf of the Senate and the College he was ashamed of me; and that he regarded the situation such that my colleagues ought to feel it a disgrace for me to continue to be a teacher, and should take action for my removal.

I was aghast at the impudence of this bold attack from my own children; I had no idea what I should do, or say, or reply. How could I interpret these statements?

Now, however, it is clear that this so shameless and uncivil act had sprung from that nest of intrigue whence other similar plots were hatched; for within a few days a second letter appeared, signed by Fioravanti. This was the tenor of it: *That he was ashamed of me for the sake of his country, the college, and the faculty; that the rumor was being circulated everywhere that I was using my boys for immoral purposes; and that not satisfied with one, I had added another to my household—a state of affairs absolutely unprecedented; he asked that in the name of many interested persons I should have regard for the notorious scandal I was causing, since nothing else was discussed in Pavia save these plans for ousting me; that houses of certain citizens where these infamies were being committed were to be publicly designated.*

When I had read these words I was dumbfounded, unable to believe that this letter was his work—the work of a friend and a sober man. Yet the memory of that former letter from my son-in-law—as I supposed—still rankled. And now I gave up all idea that it was ever his composition, that he had ever devised such a thing; since, from that time to this day, whatever his mood—amiable or vexed—he has been steadfastly devoted to me, nor ever shown a trace of any adverse attitude toward me, to say nothing of an absurdity like that letter.

I pass over further consideration of why a man in other respects prudent should, even had he believed such tales, have committed them to a letter which might fall into the hands of any number of people; how he could have charged his father-in-law with a crime, uncertain as he must have been of it at best, and a crime so filthy, so base and so certain to result seriously for him.

I called for my cloak and went to Fioravanti. I questioned him about the letter; he confessed that he had written it! Then was I even more astounded, for I had absolutely no suspicions of treachery, much less any reason for harboring such a thought. I began to urge him for his motives and to inquire where under heaven the so well-known designs for ejecting me were being concocted. At this he began to hesitate, and knew not what to respond. He could talk only vaguely about notoriety and the Rector of the Gymnasium; the latter was entirely in favor of Delfino.

Nevertheless, when the latter perceived that the situation began to look as if he himself rather might be drawn into serious difficulty than I ensnared in a suspicion of crime, he changed his plans, and withdrew; for he well realized, for all his simplicity, in what a grave situation he was involved. Accordingly, from that day all these intrigues died away, and all their well-devised inventions tottered and fell.

I may explain, in addition, since I discovered it afterwards, how the affair was concocted. The Wolf and the Fox had persuaded the Sheep[2] that if I had not been there the Senate by all means would have appointed him to a professorship at Pavia; he stood second in line. My competitor was thereupon arrogating to himself, under a certain ancient and customary tradition of succession, the right of retaining the first place as his own—Fox that he was! And it did not turn out that way as later events were witness.

The curtain having thus descended on the first act of the tragedy, the second was begun, in which some of the mysteries of the first were cleared. Above all, my enemies took care that the one who was destined to be such a disgrace to his country, his family, the Senate, the medical associations of Milan and Pavia, the faculty of professors, and finally to his pupils, should be invited to enter the Accademia degli Affidati at Pavia of which many learned theologians, two princes, the Duke of Mantua, and Marquis Pescara were already members. When my enemies observed that it did not please me to be thus drawn in, they endeavored to inspire me with apprehension. What was I to do? Disheartened by the death of my son and tried by every adversity, I at length acquiesced, particularly since the Senate would free me within a period, already definitely determined, from my duty of teaching in the Pavian Gymnasium. Not even then was I aware of their trick, nor thought it strange that they should desire association with him whom all the powers not fifteen days earlier had wished to proscribe—the monstrous spouse of all young boys. . . . O faith of God! O savage hearts of mortals! O bitter hatred of infamous and false friends! O shameless cruelty, more fell than many serpents!

What more did they devise? In the very entrance of the halls of the Accademia degli Affidati, I noticed a beam so placed that it would be easy for it to fall and kill anyone entering heedlessly. Whether this was done by chance or by purpose I know not; but it is certain that I invented excuses for appearing as seldom as possible at the Accademia, unless unseasonably or unexpectedly, like a mouse on the look-out for a trap. But nothing came of it, whether because they deemed it unwise to perpetrate a foul deed so publicly, or because they had intended nothing at all by it, or because they had abandoned it for devising other schemes.

For one thing, when, within a few days, I was called to attend the sick son of Pietro Marco Trono, a surgeon, they had hoisted above the doorway a lead weight which had been arranged to give the appearance of a device for holding the reed-curtains back. However it may have been, or how, or by what art it was so suspended that it would fall, I never investigated. But it fell, and had it touched me, all would have been over with me. By what a narrow margin I escaped only God knows. From this time on I began to be filled with vague apprehensions, not knowing exactly what I had to fear, so greatly was I confounded.

Now give ear to the third act wherein occurs the dénouement. A short time had elapsed when that Sheep—Fioravanti—came to me asking whether I would be willing to permit the two boy musicians who lived with me to take a part in the celebration attending the singing of a new Mass. Now my tormentors, knowing that these boys were my cupbearers and pregustators, had made arrangements with my maid-servant to give me poison. Shortly before this, in truth, they had asked Ercole whether he would be interested in taking a part in the celebration; and he, suspecting no evil, had promised; but when he saw that the boys also were being invited, he began to feel uneasy. Therefore he replied that only one of the boys was a musician, not two. Fioravanti, accordingly, being a clumsy man and carried away with his zeal for getting the boys out of the house, said, "Give me both of them, for we know that he is a musician, and although he may be untrained, he will help with the others to fill the chorus of boys." Ercole replied to the two who had come, "Permit me, sirs, to talk it over with the master."

Then he came in to me and exposed their designs, with an insight so clever, that, if I had not been a lunatic or a dolt, I should easily have divined what the villains were about, although until then I had not noticed. Upon Ercole Visconte's advice, and after he had insisted that the one boy did not know even a note of music, I resolved not to let the boys go.

Not two weeks later—perhaps a little longer—these same men had returned, and inquired whether I would like to give those boys permission to take a part in a comedy. It was at this time that Ercole hurried to me, saying, "Now the story is out; they wish to remove all your servants from your table in order to destroy you with poison. Not only must you keep alert and on the look-out for treachery of this sort, but you must look into every occasion, for there is no doubt that they are seeking earnestly for your destruction."

I said, "I think as much." And yet I was not able to bring my mind to accept such a notion. "What reply shall I make?" I continued to Ercole. "Simply tell them that you cannot spare your servants." I did so, and they went away.

At length, I made guess that, after much deliberation, it was decided to do away with me entirely.

On a Saturday—June 6th, if I am not mistaken—I was suddenly awakened from sleep near midnight. I found that a ring in which there was a setting of sapphire, was missing from my finger. I called the servant to rise and search, but he sought in vain. I myself rose and ordered him to fetch a light. He went to obey, but returned saying that there was no fire. I chided him severely, and ordered him to look again. Whereupon he came back, joyfully bearing fire in the tongs, or at least an ember which had remained red hot because of its size. Since he said there was no other, I said it would do; I ordered him to blow on it. He blew three times, but seeing no hope of getting flame, was taking the candle away from the coal when a tongue of flame leaped out and lighted the candle.

"Did you notice that, Giacomo Antonio?" I cried, calling the lad by his full name.

"Indeed I did," he responded.

"What exactly happened?" I urged.

"The candle took light, although the coal gave off no flame!"

"Take care then," said I, "lest it be extinguished. Let us seek the ring."

It was found on the floor directly under the middle of the bed, a point which it could scarcely have reached unless it had been flung against the wall with great force, and bounded back.[3]

I vowed that I would not set foot out of the house on the following day. Occasion favored my vow, for it was a holy day and I had no patient. But in the morning four or five of my pupils, accompanied by Zaffiro, appeared and invited me to be present at a supper at which all the professors of the Gymnasium and members of high rank in the Accademia were to be guests. I explained that I was unable; but they thought I was unwilling to make part of the company because, as they knew, I did not dine formally. So they said, "For your sake we have postponed the banquet to the supper hour."

Again I asserted, "I can by no means accept."

They asked my reason, and I told them of the portent and the vow, so that all were astonished. But two of them kept exchanging glances, and asked several times whether I would not be loath to mar the pleasure of the banquet by my absence. I replied that I would rather do that than break my vow. An hour later, they returned and pressed me somewhat more urgently; but I persisted in my refusal; for had I not decided not to leave the house at all? However, during the evening, although the sky was overcast, I went to attend a poor victualler who was sick, since my vow did not forbid a professional call.

Thus I continued to live, vaguely confused until I went away from Pavia. Immediately it was discovered that the Fox, frisking and gloating because this had befallen, was promoted to the professorship by ruling of the Senate. But alas for mortal hopes! Not three or four times had he lectured before a disease attacked him which, according to the story which came to my ears, lasted about three months; and he died. He was utterly apprehensive because of his crime; for I understood afterwards that a certain one of his associates, as a material witness of his perfidious conduct, had been destined for a poison potion at a banquet.

Delfino died in that same year and Fioravanti shortly after. Nay, even in the machinations forged for my torment at Bologna, a similar fate overtook a like number of physicians, though somewhat later. Thus all who sought my life perished.

If, however, God had permitted me to be afflicted with so many calamities as a condition of the benefits which he has so unremittingly bestowed upon the human race, thus would my enemies have paid the reckoning. I had learned, moreover to keep at a distance from any chances of this sort, profiting by the fate of my uncle, Paolo, who had died of poisoning, and of my father who had twice drunk poison, and who, though he escaped, had lost his teeth.

Close upon the foregoing agitations followed I know not what a host of troubles. In the month of July I was forced to undertake a journey on account of a serious ailment with which my infant grandson was suffering at Pavia, while I was at Milan. From the exertions of travel I fell ill of erysipelas, which settled in my face; this, followed by toothache, put me on the verge of having myself bled, when the approaching new moon caused me to refrain. From the day of the new moon, I deemed myself much improved, and so escaped both the peril of death and the pain of treatment. Subsequently a servant threatened murder in an argument over some money; I forestalled his act only by a few hours. In the wake of these troubles a long and painful attack of gout bore down upon me.

From the year 1572, I have suffered no slight risks of my life from lurking perils, since the streets of Rome are little short of a wilderness to me, and the behavior here so uncouth that many physicians more wary than I, and far more adroit in adjusting themselves to the customs, have found hereabout the origin of their deaths.[4]

Accordingly, when I observed how I was protected more by Divine Providence than by any wit of my own, I ceased to exercise any further anxiety for my safety in dangers. Who does not now perceive that all these things have been, as it were, precursors of bliss about to be overtaken—a nightwatch waiting for the dawn; just such a period as that which, in 1562, was followed by the professorship at Bologna, a post I held eight years, an honorable and useful work with a certain intermission from so many molestations and labor, and accompanied by a somewhat more pleasant existence.

31.
HAPPINESS

ALTHOUGH the name of happiness suggests a notion far from applicable to my nature, nevertheless I have chanced to attain in part, and have had a share in, that which is indeed very near felicity.

In the first place I think I may lay a clear claim to happiness in that all events of my life have come and passed in an orderly fashion, as if by rule—if ever to any mortal such a lot may fall. Had this not been the case and had the numerous commencements of the succession of events begun a little too late, or a little too soon, or had the conclusion been delayed, my whole career would have been subverted.

Second, my happiness in any given period of time is merely comparative in relation to the whole; for instance, when I was living in the town of Sacco. Since among the giants one must necessarily be the smallest, and among the pigmies one must be the tallest, yet the giant is by no means little, nor yet the pigmy large; so, although while I was in Sacco I was comparatively happy, it does not follow that I was of a happy nature. I gambled, I entertained myself with music, I rambled about the country, I feasted, and, though I accomplished but little through devotion to my studies, yet I was never apprehensive, having neither anxieties nor fears. I was respected and not without opportunity to enjoy social intercourse with members of the Venetian nobility. It was the very flower of all my time; nothing in my life has been happier than this pleasant existence which lasted five years and a half, from the month of September in 1526, to the month of

February 1532. I used to exchange views with the mayor; the public house was a kingdom and a rostrum. That pleasant dreams now carry me back to those days is evidence of my erstwhile happiness which sprang from my sense of enjoyment, even though that time has long since vanished and the very memory of it shrunk to a slender thread.

My third observation is more important than the others; for just as it makes for happiness to be what you can, when you cannot become what you would, so it makes for a more abundant happiness to be desirous of that which is best among all those things we long for. It is necessary, then, that we should recognize what we have at our command, and make the most expedient selection of such faculties as are best for our purposes—choosing two or three from the number which are of a kind to claim our ardent affection and desire, so that we may possess them with the least possible disadvantage both for the purposes for which they were set apart and for other purposes as well. Finally, it is fitting that what we claim to possess, we possess with the best possible right. To possess is one thing; to possess what is best is another thing; while to hold unchallenged right to what one has is the perfection of possession.

I do not overlook the fact that some who contemptuously regard these things as paradoxes will fail to agree; but to anyone contemplating the vanity of mortal concerns, and reviewing experience, it will be easy to understand that such things are far truer than we would like to have them. But if you should continue to disagree, reflect that time brings all things to light, and shows what estimate it puts upon them. Let us take, for the sake of example, Augustus, Marcus Scaurus, Seneca, and Acilius.[1]

Certainly Augustus was a man of abundant good fortune, and, in the judgment of men, happy. What now remains of him? His very stem and stock has passed out of existence, his very memorials: all things Augustan, in short, have ceased to be—

nay, his very bones. Who even wishes to be connected with him, to say nothing of being eager? Who becomes angry with anyone offering violence to his name? And were it even otherwise, what is it to him? Was he happy while he lived? What more did he possess than his other associates of blessed memory, beyond the anxiety of constant thinking, wrath, madness, fears, and murder? His house was full of tumult, his court rife with confusion, his intimates plotting against him. If he could not sleep he was miserable; if sleep he did, slumber overcame his vigilance, and vigilance is untrustworthy when sleep, which is heedless of mortal anxieties, is the stronger.

There was Marcus Scaurus—what true felicity had he with all his great treasuries, with his public entertainments and lavish displays, of all of which not even a shadow of recollection remains? In his lifetime did he not experience confusions, anxieties, the wakeful restlessness and toils of authors? Did not the great shows, the pomp and voluptuous pleasures of others afflict him?

What fair fortune attended Acilius when all his resources were swept away, when pain had advanced from below to take the place of pleasure, and dire poverty the place of riches? I shall not labor to show that Acilius was wretched; there is no stronger surety for an utterly unhappy life than to be reduced to misery after leading in bygone days a joyful and prosperous existence.

What of Seneca's heyday? Naught persists save an evil rumor. His soul cannot be happy even though he lived with a following of reeking youths, ate at tables of cedarwood and of ivory, received interest on his profitable investments from all Italy, and possessed so many gardens that their mention has passed to a very adage. Afterwards he was forced, because of fear of the poisoners of Nero, to live on bread and fruit alone, with water drawn from the fountains. In after years but a fraction of his great possessions sufficed a grandson accustomed to the very quintessence of luxury: all over and above contributed to excess, not to living, a fact which is evidence that the essentials of a happy existence are not actually lacking to any man.

What further confirmation of my doctrine does the foolish pomposity of Lucius Sulla contribute? Who, after Marius had been slain, ordered himself to be called Felix, when he was more than ever a contemptible old man, defiled with murders and proscriptions, with a host of enemies, and without the magisterial powers which he had cast aside. Wherefore, if it is legitimate to be happy under such conditions, I, lacking these, am so much the more worthy of that name.

Let us live, therefore, cheerfully, although there be no lasting joy in mortal things, whose substance is evanescent, inane, and vacuous. But if there is any good thing by which you would adorn this stage of life, we have not of such been cheated—rest, serenity, modesty, self-restraint, orderliness, change, fun, entertainment, society, temperance, sleep, food, drink, riding, sailing, walking, keeping abreast of events, meditation, contemplation, education, piety, marriage, feasting, the satisfaction of recalling an orderly disposition of the past, cleanliness, water, fire, listening to music, looking at all about one, talks, stories, history, liberty, continence, little birds, puppies, cats, consolation of death, and the common flux of time, fate, and fortune, over the afflicted and the favored alike. There is a good in the hope for things beyond all hope; good in the exercise of some art in which one is skilled; good in meditating upon the manifold transmutation of all nature and upon the magnitude of Earth.

What, then, is evil in the midst of so great an array of blessings and such a store of wisdom, that the universe abounds in hope! We do not actually live unhappily; and unless the general condition of mankind should be inconsistent with it, I should dare to predicate abundant happiness—but an unseemly and futile design it is to beguile others into a deceptive expectation.

If free choice of place should be given me, then, I should move my abode to Aquileia or Porto Venere as delightful places to live; or outside Italy, I should choose Eryx in Sicily or Dieppe on the Drudis River,[2] or Tempe's Vale in Thessaly. For my age does not carry me as far as Cyrene's shores, nor to Mount Zion in Judea; nor does it urge me to seek Ceylon's Isle near India. These regions may bear happy men; they cannot make men happy.

32.
HONORS CONFERRED

FOR HONORS I have neither been eager nor yet consumed with longings; neither was I much their lover, since I realized how much grief they brought into human existence. Wrath is a great evil, but it passes; the misfortunes of high rank are never ending. First, a zeal for honors drains away our resources while we evade laborious effort; we exhaust remaining opportunities of gain while we spend lavishly for clothing, give rich banquets, and support a retinue of servants. Again, a zeal for honors urges us to the verge of death itself by ways too numerous to recall—duels, wars, quarrels, disgraceful litigation, attendance upon the favor of princes, indecorous feastings, intercourse with wife or woman.

We challenge the seas, we profess to think it the honorable thing to fight for one's country. It was a characteristic of the Brutus family to devote themselves to death in battle; Scævola burned his own right hand; Fabricius rejected the bribe.[1] The last was perhaps the act of a prudent man; the others were the deeds of foolish, not to say mad, men.

There is no reason why one should vaunt his country. What is one's country save a cabal of petty tyrants for the oppressing of the weak, the timid, and those who are generally innocent? I refer particularly to the Romans, the Carthaginians, the Spartans, and the Athenians, who being unscrupulous and wealthy wished, under the pretext of such an agreement, to lord it over the honest and the poor. Oh, shameless mortals! You surely do not think those men so prodigal of life that they long to suffer death for their country for honor's sake!

The poor rustic, on the other hand, argues thus with himself: "Alas, what condition is more wretched than mine! But if I survive the struggle I may be numbered among the chiefs of the state; even as this or that man, I shall possess myself of the goods of others; yes, even as others now lay claim to my possessions. If I die, my descendants will be called from the plough to the chariot, and to triumphal pomp!" This was his love of country, his spur to honor!

I do not wish to make subject to these indictments those cities which fight for civic liberty, nor those princes the sole guerdon of whose principate is to watch that justice be done, to encourage good men, to lighten the burdens of the poor, to cultivate the virtues, to reform their own connections, and to check those who have already obtained positions of great power. It is not a blemish on a city to wish to have dominion, nor a crime for the citizens to wish to emulate the natives of other cities; the very necessity of order brings about domination, and want of rectitude the despot's sway.

But let us return to the commonplace, for but few have true understanding of the things I have been discussing. Honors filch away time, while we are forced to give ear to so many men, to so many salutations. A desire for honors steals away our devotion to wisdom, man's one divine endowment. Oh, how many mortals are leading unprofitable lives—a heedless audience! Who if they would expend in study those two hours which they consume each day in combing and adorning themselves, in the course of two or three years might be able to speak authoritatively on some subject! Yet this very triviality is the prize and the dignity of honorable rank!

Honors likewise interrupt the guidance of a family and of sons—and what is a greater folly? Honors subject us to envy; from envy spring hatred and rivalry, and from these, in turn, disrepute in the eyes of the people, contamination, accusations, deaths, loss of property, retribution. When such disasters have come to pass we lose a great part of our freedom, if not the whole; and (if one is of a mind to agree with the popular sentiment) we lose whatever good is left in life—voluptuous pleasure itself. Along this path we advance, under honor's auspices, to the most dangerous of evils—a craven spirit. Impudent youths enter upon this course, imperilling, thereupon, the very integrity of homes, and ruining them.

All things men call good possess the essence of good, except ambition. Sons bring with them the hope of succession; friendship forms no slight share in happiness; resources guarantee us all comforts; virtue becomes a solace in misfortune and a distinction in felicity; associations and guilds are a great source of security, now, and always at other times. Ought we, consequently, strive to avoid high and lasting preferments, and to flee them?

Certainly not! Primarily, since public honors increase one's means, power, and profit, as in the case of officials, physicians, painters, and in the arts in general; whence it is customarily said that honors foster the arts. In the second place, honorable rank relieves a man from perils, as a guild or association exerts itself in behalf of those who labor in the toils of harsh malice, in false calumnies, and preposterous charges. In the third place, honorable position adds power, and this is notably the case in the military profession and in certain magistracies, wherein the advancement rests upon an appointment to the office of adjutant, as they say, and thence through the successive degrees in order of rank. In general, a place of honor lends power for repairing conduct worn by ignominy and disgrace, which would appear the more indecorous by comparison with the honor; again, a place of honor lends power because of the very advantages which accompany the circumstances related to it; it is, besides, of benefit when one is engaged in places where he is not known. A fourth point is this: occasionally one may, without much trouble, make his honors do service in lieu of talent.

In favors of this sort, then, chance has not been grudging: as if in augury of my fortune, many used to offer me, Cardano's boy, companionship, whenever I appeared where he was, and when I acquired some skill in Latin. I was known at once throughout my own city. Later when I went to Pavia, many sought my friendship voluntarily; and an opportunity was presented, which, had I embraced it, would have opened the way to preferment, sooner or later, from Pope Pius IV. I debated with reasonable success, and Matteo Corti challenged me, *honoris causa.* I held a professorship for mathematics for one year only, as the academy was dissolved in 1536, and I was summoned to Pope Paul III. Then in 1546, after I had already been lecturing for the two years previous on mathematics—geometry, arithmetic, astrology, and engineering—an occasion arose, through the instrumentality of the great Cardinal Morone, of lecturing at Rome with an honorarium from the Pope himself.

In the following year, through Andreas Vesalius and the Danish Ambassador came a proposal, with a consideration of three hundred gold crowns yearly in Hungarian money, and six hundred in addition in impost-money, derived from the tax on skins of value. The latter revenues differ from the royal currency by an eighth part, and are collected somewhat more slowly; they are not even redeemable at par, and are in a measure subject to certain hazard. The consideration included living for myself, five body servants, and three horses. A third call came from Scotland. I blush to name the emolument offered, but it is enough to say that in a few years I should have been a very wealthy man.

The first offer I rejected because the climate in that region was exceptionally cold and damp; the natives were clearly uncultivated; the ritual and doctrines of their religious observances were greatly at variance with those of the Roman

Church. The second I also rejected because it was not permitted to effect an exchange of moneys through brokers and couriers into England, much less France or Italy. This was in 1552, in August.

In the month of October the Prince of Bois-Dauphin and a confidential agent of the King of France,[2] Vilander, came to me with another proposition amounting to eight hundred crowns annually; and on the sole condition that I should be willing to kiss the royal hand, as the saying goes, and be off immediately[3]; they included in their terms a heavy chain worth five hundred gold crowns as compensation. Other agents were soliciting me in behalf of the Emperor who was besieging Metz. I was unwilling to come to an agreement with either party; with the latter because I realized that excessive hardships were involved in accepting the post—the Emperor had already lost the greater part of his army through cold and hunger; with the former because it was not meet that I should leave the domain of my sovereign, and transfer my allegiance to his enemy.

When I journeyed from Antwerp to Basle, the illustrious Carolo Affaidato invited me to stay at his country home, and, quite against my wishes, tried his best to present me with an excellent mule, the value of which almost exceeded an hundred crowns. Affaidato was a singularly cultivated gentleman, very generous, and a lover of men known for their high character.

On the same journey, a Genovese noble, Ezzelin, offered me an Asturian riding-horse; the English in their tongue call such an horse "Obin."[4] A certain delicacy prevented me from accepting this gift; otherwise, I should have taken it, since nothing, according to my taste, more beautiful or more magnificent could have been desired; the creature was all white and sleek. There were, moreover, two horses exactly alike, and Ezzelin offered me my choice.

The next year, Prince Don Ferrante,[5] as he was customarily called, offered me a consideration of thirty thousand crowns, to enter for life the service of his brother, the Lord of Mantua, as they call the Duke. They were ready to discharge a thousand crowns, among other things, on the first day; but it did not please me to accept. Don Ferrante was amazed and offended. He was the more disposed to this attitude inasmuch as he had approached me with a proposition honorable, according to his standard, but to me not so. His persuasions were of no avail, and he was reduced to threats. Finally he came to understand why I preferred, like the ermine, to die rather than to be stained; and he had much greater respect for me from that day forward, an usual characteristic of men of magnanimous spirit.

The sixth opportunity for preferment came in 1552 from Viceroy Brissac, who offered all sorts of inducements, influenced as he was by Lodovico Birague, my illustrious fellow-townsman. Others, who had sought my services, had done so, desiring me in the capacity of a physician; Brissac wanted me as an engineer, but that was far removed from my wish.

But before I turn away from this subject, I wish to show why so rich an emolument was offered by the King of Scotland, a very poor monarch, who has, according to report, an income of not more than forty thousand gold crowns annually. The chiefs—they say to the number of fourteen thousand, although I do not think there are that many—under allegiance to the royal power are obliged to rally to arms in camp for a period of three months, upon summons. And, when anyone of them dies, the King becomes guardian of his sons—according to an old practice to which the King has ascribed this function—until the eldest son has attained his twenty-first year, or come to his full majority. In the meanwhile, living and clothing excepted, all incomes are the guardian's without any account rendered, for they come under the estate of the King. And what is more, whether the ward be male or female, the guardian has the right to contract a marriage for the ward with anyone the guardian wishes, the only condition being that the man or woman to whom the ward is married must also be of noble birth. The ward is forced to accept a dowry not on his own terms, but on the terms of the King, acting as guardian.

But to return to my own affairs: thrice I was engaged at Pavia, four times summoned by the Senate of Milan, and three times by the Senate at Bologna, although the last engagement was without advantage.

On my return from Scotland, forty men of rank awaited me at Paris in order that they might gain some knowledge of my methods in the art of medicine, and one made me an offer of a thousand crowns merely on his own account. But I was scarcely able to cross France; indeed, I dared not.[6] Concerning this incident I have a letter from a high Parisian officer, Aimar de Ranconet, a man, as I have elsewhere noted, well versed in the classics. While I traveled from France to Germany I was received everywhere much as Plato was long ago at the Olympic games.

I have repeated the testimonial, also, already mentioned, of the College at Padua to the Prefect of the city. An honor not unlike this came to me in Venice. It was an honor which could not be conferred unless all voting upon it were unanimously agreed. There were sixty votes; I obtained them all![7] I had similar good fortune when I was elected by the Senate of the College at Bologna. There were twenty-nine votes; I received twenty-seven, when by so small a margin as twenty-five I could not have succeeded. Not only did my success reach every ear, but my name also was known to princes, kings, and emperors of the whole earth. And this cannot be denied, although it may seem too foolish and vain to mention, that even if one or another of these honors is made light of, the sum of them cannot be scorned as unworthy of regard: the knowledge of many facts, travel, acquaintance with danger, the offices which I have held, the offers of preferment which have come to me, the friendship of princes, my reputation, my books, remarkable cases of healing, and among other circumstances which came to pass, certain singular characteristics of a supernatural cast, a guardian spirit[8] besides, and that understanding born of a flash of intuition. I was a member of three Colleges of Physicians—at Milan, at Pavia, and at Rome. Of all these honors, moreover, I sought none, by so much as a word, excepting my admission to the College of Physicians at Milan, into which I was admitted by election toward the end of August 1539, and my professorship at the University of Bologna. In both cases I was forced by necessity, not spurred by ambition. The privilege of citizenship was also conferred upon me by the Senate of Bologna.

In general, as I have said, desire for high honors is a disastrous ambition to mortals, but inasmuch as I am considering the matter of honors, let me say that I have attained not merely something more, but indeed, much more than I had hoped, as, for instance, that title[9] [*Vir Inventionum*]. But let us take up a discussion of dishonor.

33.

DISHONORS. WHAT PLACE TO DREAMS? THE SWALLOW IN MY COAT OF ARMS

ONE OUGHT to construe the question of dishonor in terms vastly different from those of honor. It is, for instance, better for the majority to flee preferments; yet, of a truth, it is the accepted belief that to tolerate disesteem is advantageous to none, nor even respectable. There is an old saying: "Veterem ferendo iniuriam invitas novam."[1]

For when one admits dishonors they will attack and ruin his career as a thing of no account; nor is this a concern merely of man's early years, for boyhood is not open to disgrace especially if one adopts that Horatian ideal regarding a parent's part in the education of the child:

> Ipse mihi custos incorruptissimus omnes circum
> doctores aderat.[2]

Even as I have said, then, the toleration of disgraces and baseness begets, first of all, every detriment which honors breed, only greater. Such, especially, are the dealings of men in circumstances when women are concerned. Since, verily, dishonors as well as honors are double-faced, that fate of man is truly calamitous which depends on superficial honors. Inwardly, though, his lot is blighted with shameful conduct; he is acclaimed by fools, or by the crowd; superior men hold him in odium and the insolent ridicule him.

On the contrary, those whose honors are not in evidence are lightly esteemed by society; their existence is safe and tranquil if they permit it to be, and in so far as they are contented with their condition. Those who are slighted and lacking on both counts are as the laborers and tillers in our midst, who if they live under just princes flourish well enough, especially if they unite their interests through a society. Those who for a surety enjoy both the inward and outward evidences of honors are acquainted with the bitters of calumny and with secret machinations; they stand greatly in the way of danger from accusations; from public trials they are safe because people stand in fear of hatred or danger to their lives in offending men so privileged.

While I was at Bologna, when an agreement about my salary was still pending, two or three persons came to me by night and in the name of senators and judges sought that I should put my name to a petition to release a woman already condemned by the authority of the city magistrate and the pontifical power for impiety, magic, or sorcery. They tried to persuade me by reasoning that according to the philosophers there are no demons. They wished me to order freed from prison another woman, this one not yet sentenced by the judges, on the grounds that she was sick to the point of death under the treatment of other physicians. Besides this, they presented packages of horoscopes asking that I should make some pronouncement upon these as if I were a soothsayer and prophet, and not in my capacity as professor of medicine. These men, nevertheless, got nothing for their pains save their fruitless endeavors, and an unsavory reputation with me.

However, to begin with, when I was twelve years old, and had discharged a rocket full of gunpowder, and the respectable wife of a musician had been hurt by a piece of the paper covering, I was struck so sharp a blow on my jaw that I reeled.

A debate, held at Milan, was undertaken with insufficient powers on my part, and turned out with little credit to me. Because of the violence of certain doctors, I was forced, in 1536 or 1537 to come to dishonorable terms of recognition with the College of Physicians; but, as I have elsewhere explained, sometime after, in 1539, this condition was revoked and I was restored to my full rights.

In the year 1536, while I was attendant physician to the house of Borromei, I had a dream early one morning about a serpent of huge proportions by which I feared, in my dream, I would be destroyed. Not long after a messenger appeared to summon my services for the son of the so-called Count Camillo Borromeo, a man of note and rank. I went thither. The boy, who was seven years of age, seemed to be afflicted with some slight illness, but his pulse, according to my observations, regularly missed every fourth beat. His mother, the Countess Corona, asked me how the child was. I replied that he seemed to have but little fever, yet because the pulse failed at every fourth beat, I knew not exactly what to think of it. I did not have at that time Galen's books on the Indications of the Pulses. When all his symptoms had remained the same for three days, I directed accordingly that he be given to drink small doses of a medicine which is called *Diarob cum Turbit.*[3] When I had written the prescription, and the messenger was already on his way to the pharmacist's, I remembered my dream.

"How do I know," I questioned myself, "whether this lad may not die according to the symptom indicated?" And after the books were available it was found to be so. "These consulting physicians who are so hostile to me will lay the cause of his death to the medicine."

I called back the messenger who was not yet four paces from the door, and said there was some certain ingredient lacking which I wished to include in the prescription. Cautiously I tore up what I had first made out, and wrote another prescription of pearls, bone of unicorn,[4] and gems.

The powder was given, and the boy vomited; all present knew his case was serious. Three leading physicians were summoned, but one of these was less unbending than the rest—he who had been present in the case of Sfondrati's son.

They saw the description of medicines, and asked, "What are you doing for him?" Although two of these hated me exceedingly, since it was not the will of God for me to be further assailed, they not only praised the medicine, but ordered it to be administered again. This done, I was saved.

When I had visited the patient again at evening, I knew all; on the following morning I was summoned again at break of day. I saw the child was laboring in his last agonies and the father overcome with tears.

"Look at him," he cried, "you who said he was not seriously ill." As if I had said this! "At least you will not go away while he still lives!"

I promised to stay. Thereupon I observed that the father was being restrained by two noblemen, and as, crying out, he attempted to rise, they would not let him; yet he kept reiterating that I was to blame. What more? If there had been any mention in the prescription of *Diarob cum Turbit*, which in truth was not safe, it would have been all over with me, since while the man lived he would have condemned me so relentlessly that no one would have sought my services: as if Canidia would have breathed her worse-than-African-serpent breath upon me.[5]

Thus I escaped death, and transferred my loss from disgrace into zeal, lest in the future I should have occasion for regret.

I shall not believe that dream, as well as other occurrences narrated above, to have been fortuitous; it is plain to be seen that they were warnings, to be taken as such by a pious soul oppressed by many afflictions, whom God was not willing to forsake. But—although the vision fitted the occasion—I was excited by this stimulus: there was a house in the square of Santa Maria Podone entirely decorated with painted serpents after the fashion of crests which have joined adders as in the ancient coat-of-arms of the Borromei.

We Cardani had for a crest a red castle with its towers; but the high turret in the midst is black, and the whole is against a white ground, so that the turret may be distinguished from that of the Castiglioni, whose illustrious name calls for a walled town with a lion. The Emperor, accordingly, added to the coat-of-arms of the Cardani an eagle, entirely black, except the beak, and with wings outspread upon a saffron field; some branches of the family depict the eagle with a single head, and others with a double or divided head. Likewise, certain members of the family color both fields saffron, while others, following the ancient practice, make the ground of the eagle saffron and the ground of the castle white. But I, from the day of my imprisonment, added to my seal the image of a swallow singing under the eaves of a barn.

To reproduce these symbols was a most difficult matter on account of the variety of colors. I chose the swallow as in harmony, in a most extraordinary way, with my own nature: it is harmless to mankind; it does not shun association with the lowly, and is ever in contact with humankind without becoming familiar. It changes its habitation with frequent goings and returnings; it is connubial, not solitary, yet not disposed to flock with others. It delights its host with a song; it ill endures the cage. To the swallow alone, if it be very young, eyes return after they have once been plucked out. The swallow bears in its belly certain exquisite gems.[6] It takes marvelous delight in the serenity and warmth of the skies; exceptionally skillful in constructing its nest, it yields honors in this art to the kingfisher alone among all the birds. Though white-breasted and white within the wing, the bird is dark on its outer parts. As if grateful and mindful, it returns to a hospitable home. No other bird, not even the bird of prey, pursues it; not one surpasses, much less equals it, in flight.

But to return to my story. On two occasions in the town of Sacco, an attempt to cure turned out disastrously for me. A woman in the town, Rigona by name, from whose foot I had let blood on the sixth day, died on the seventh. Another patient, who was a poor fellow and who, as a *patronalis* of the church, as they call it, was accustomed to ring the bell or the *campana*, passed away that very night if I remember correctly, after I had given him medicine. The cause was the same in either case: because the disease with which each labored was a malady I did not know. I was, in truth, on the verge of a disrepute which, had it increased, would have been my downfall and the ruin of all my career, private as well as public.

The case of Vignano also turned out unfortunately—yet this side of death—and for this he withdrew from my care, although nine from his household had been restored to health through my efforts.

Thrice, therefore, in the fifty-one years in which I have been practicing medicine, I have been in error; while Galen did not reproach himself as often as he erred, on the ground that the reproach is no extenuation for the mistake. And so I have never been exposed to any public disrepute; there have, to be sure, been occasional accusations with no confirmation, specially at Milan and at Bologna—but nothing more; and I believe that they, there, would not have been so ungracious to a fellow-townsman. At Pavia, however, there was nothing of this sort, although I was indeed a citizen of that town, having been born in it and having found a residence there.

Truly, as fever ends certain diseases and frees those sick with it, who could not otherwise be healed, from death, according to the teaching of Hippocrates in his *Aphorisms*,[7] so prison, overwhelming the shoutings about my so many crimes, cancelled all, in such a way that no gleam of suspicion appeared to reveal how potent a force envy may be.

And you physicians: What have I done to you? All their infamies, thereafter, all their vituperation, found, as it were, an end in the event which they had hoped to make serve as a beginning.

But let us lay aside these harangues about infamy and

disrepute which at our age have more to do with the scandals of womenkind than of men.

Was that not a remarkable fabrication at Bologna of a Dialog which was circulated in my name under the title *Melanphronis*—that is, in the name of recondite and black knowledge? But so crudely was it framed and so unhappily published, that the very trial of it showed that the author of the invention had no knowledge at all, either white or black! Consequently the publishers themselves were forced to destroy their work.

But let us return to my own country where I was rejected as professional attendant to the Hospital of St. Ambrose, a post which offered a yearly stipend of seven, not quite eight, crowns; and this occurred, if I am not mistaken, in my thirty-seventh year. Already, long before this, I had failed in my attempt, in my twenty-ninth year, to secure the practice in the village or town of Caravaggio, where the annual income did not reach eighty crowns—and regular horse's work at that!

At Magenta an agreement had already been contracted with me. The stipulation was in the neighborhood of fifty-five crowns; but on the very eve of settling there, I withdrew, and so barely escaped rotting away in the place, not to say that I could ever grow old there.

Also, in my twenty-ninth year, although some friends were willing to intercede on my behalf, I did not accept the post of physician at Bassano with annual returns amounting to an hundred crowns. This town lies in the district of Padua.

It was the gratuitous and able advice of Cesare Rincio, one of the foremost medical men of our State, that I ought to be willing to take up a practice, at a stipend of twelve crowns a year, in a certain village in the district of Novara, fifty miles away from my own city of Milan!

You will not be surprised then if I say that in the village of Sacco my fees amounted to nothing a year, and I persevered five years. Moreover, Giampietro Pocobello at Mozzate and

Giampietro Albuzzio at Gallarate were bequeathed incomes of about twenty crowns a year; this was the result of having married in the hope of an inheritance, although neither of them ever contracted another marriage, since the first wife survived in each case.

34.
MY TEACHERS

MY FATHER, in my earliest childhood, taught me the rudiments of arithmetic, and about that time made me acquainted with the arcana; whence he had come by this learning I know not. This was about my ninth year. Shortly after, he instructed me in the elements of the astrology of Arabia, meanwhile trying to instill in me some system of theory for memorizing, for I had been but poorly endowed with ability to remember. After I was twelve years old he taught me the first six books of Euclid, but in such a manner that he expended no effort on such parts as I was able to understand by myself.

This is the knowledge I was able to acquire and learn without any elementary schooling, and without a knowledge of the Latin tongue.[1]

I was almost twenty when I entered the University at Pavia. Towards the end of my twenty-first year, I held public debate with Corti, one of the heads of the faculty of medicine, who deigned to honor me by holding disputation with me, a thing for which I had by no means dared to hope from him. I attended the lectures of Branda Porro in philosophy, and at times listened to Francesco Teggo of Novara. In the year 1524, I again heard Corti at Padua, and Memoria in Medicine. At the same time, I saw Girolamo Accorombone in his capacity of Professor of the Practice of Medicine, as they say, while on the faculty were Toseto Momo and a Spanish philosopher of great name.

35.

WARDS AND STUDENTS

THE FIRST of my pupils was Ambrogio Bigioggero who became pilot of a ship out of Epidauros. He was an ingenious and daring man. The second, Lodovico Ferrari[1] of Bologna, who later became professor of mathematics at Milan and in his own city, excelled as a youth all my pupils by the high degree of his learning. Giambattista Boscano was a third pupil. He became receiver of petitions of the Senate of the Emperor Charles V. A fourth was Gasparo Cardano, grandson of another Gasparo, a cousin of mine. This youth became a physician and lectured publicly on medicine at Rome. A fifth pupil was Fabrizio Bozzi who took charge of the militia in the region of Turin although considered a Milanese. Giuseppi Amati, private secretary of the governor of our province, was a sixth. A seventh, Cristofero Sacco, was made notary public; while the eighth, Ercole Visconti, a charming and agreeable young man, was a musician. My ninth pupil was Benedetto Cattanei of Pavia who entered upon the practice of jurisprudence. Giampaolo Eufomia, the tenth, was a musician, and a man of no little erudition. Rodolpho Selvatico of Bologna became a physician, and even as I write these words, he is practicing medicine here at Rome. The twelfth of my pupils was Giulio Pozzo, a native of Bologna. This one alone failed to fulfill his contract. A thirteenth, Camillo Zanolini, likewise a Bolognese and a musician, became a notary public, and a man exceptional in the elegance of his manners. Ottavio Pitti, from Calabria, and with me at present, is my fourteenth pupil.

Of all these the most outstanding are the second, the

fourth, and the eleventh; the first two, however, died while yet young; the second in his forty-third year, and the fourth when he had not yet entered his fortieth year.

> Immodicis brevis est ætas, et rara senectus;
> Quicquid amas cupias, non placuisse nimis.[2]

36.
MY WILL

I HAVE DRAWN many wills up to this day, which is the first of October of the year 1576.* The last was upon the advice of Jacopo Machelli and Tommaso Barbieri both of Bologna. I made others at Milan, Bartolomeo Sormano, Girolamo Amati, and Giangiacomo Crivelli being my various advisers. And now I have decided to draw another—my last, but I have appended codicils.

The most important of all is that I should like to have my property pass to my son, if this be possible; but my younger son has shown himself a youth of such evil habits that I should prefer to have all I own pass to my grandson by my eldest son.

The second is this: that all of my descendants remain under the care of the guardian as long as it is feasible, for certain reasons known to myself alone.

The third: that my property be subject to entail, and that when my progeny shall fail, then I wish my bequest to pass to my kinsmen, in so far as this is possible, as a whole, *in perpetuum.*

The fourth clause is this: that my books be collected and published, to the end that they may prove of such use to human kind as was the intention of their composition.

The fifth clause: that the house at Bologna shall, when succession fail, become a college dedicated to the family of

---

*Thus in the original; but Cardano died September 20, 1576.

the Cardani, and any heirs who may not be in direct succession are, nevertheless, to assume the family name.

A sixth clause is added, giving authority for such changes as may be required by any conditions that may arise.

37.

CERTAIN NATURAL ECCENTRICITIES; AND MARVELS, AMONG WHICH, DREAMS

THE FIRST evidence of a nature, so to speak, anomalous, was the very fact of my being born with long, black curly hair: not, to be sure, a circumstance supernatural, but all the same, full of portent. And what is more extraordinary, I came into this world well-nigh suffocated.

The second remarkable manifestation appeared in my fourth year, and lasted about three years. According to injunctions of my father I rested quietly in my bed until the third hour of the day; and as I lay awake awaiting the striking of the customary hour, I spent whatever time there was in the presence of an agreeable visitation. Nor was I ever foiled in my play and led to expect it vainly. For I used to vision, as it were, divers images of airy nothingness of body. They seemed to consist of very small rings such as compose vests of chain mail—although up to that time I had not yet seen a linked cuirass. These images arose from the lower right-hand corner of the bed, and, moving upward in a semicircle, gently descended on the left and straightway disappeared. They were images of castles, of houses, of animals, of horses with riders, of plants and trees, of musical instruments, and of theaters; there were images of men of divers costumes and varied dress; images of flute-players, even, with their pipes as it were, ready to play, but no voice nor sound was heard. Besides these visions, I beheld soldiers, swarming peoples, fields, and shapes like unto bodies which even to this day I recall with aversion. There were groves, forests, and other phantoms which I no longer remember; at times I could see a veritable chaos of innumerable objects rushing dizzily along *en masse*, without confusion among themselves, yet with terrific speed. These images were, moreover, transparent, but not to such a degree that it was as if they were not, nor yet so dense as to be impenetrable to the eye; rather the tiny rings were opaque and the spaces transparent.

I was not a little delighted with my vision, and gazed so wraptly upon these marvels, that my aunt on one occasion questioned me whether I saw aught. And I, though I was still a boy, took counsel with myself: "If I tell, she will be displeased at whatever causes the passage of this proud array, and she will do away with my phantom festival." Even flowers of many a variety, and four-footed creatures, and divers birds appeared in my vision; but in all this exquisitely fashioned pageant there was no color, for the creations were of air. Accordingly I, who neither in youth nor yet in my old age have been given to falsehood, stood still for a long time before I replied.

Thereupon she asked, "What makes you stare so intently?" I no more remember what I replied to her; I think I said, "Nothing."

A third peculiarity of mine was that following this period of visions, I scarcely ever, until nearly daybreak, had any warmth from my knees down.

A fourth was that all at once, out of a profound sleep, I would find myself drenched in a warm, heavy sweat.

The fifth evidence of my eccentric nature was the oft-repeated dream of a cock which I feared would address me in a human voice, and which, somewhat later, actually did this very thing. His utterances, moreover, were for the most part threatening, but I have not remembered what I heard on these many occasions. There was also a cock with crimson feathers with comb and wattles of the same hue, which I believe I saw in dreams more than one hundred times.

After epochs of this sort, when I was approaching manhood, although these visions ceased, two other illusions appeared which have remained with me almost constantly, and even now are present at times. To be sure, since I have written my *Problemata* and talked them over commonly with my friends, one of them has ceased to manifest itself so frequently: that is, that as often as I lift my eyes to heaven I see the moon; and truth to tell, I seem to see it placed directly in front and facing me, the cause of which I have explained in the treatise just mentioned.

Another curious circumstance is the following: when I had happened to observe by mere chance, that whenever I became party to a brawl no blood was spilled nor anyone wounded, I purposely took part in several combats and riots, but not a wound was dealt to anyone involved! When I joined a hunting party, not a single beast was wounded either by a hunting-spear or by the dogs; this I have taken account of as many times—though seldom enough to be sure—as I have accompanied the hunt, and have never failed to exercise this faculty.

On one occasion, indeed, when I was a member of the hunting party of Prince d'Iston to Vigevano, a hare had been run down, and, though retrieved in the teeth of the dogs, was found without a wound, to the no small surprise of those present.

Only in voluntary blood-lettings and in the cases of those publicly punished, has this, which I may call my prerogative, been powerless.

Once in the archway of the great church at Milan when some of his enemies had thrown a certain man to the ground, and had inflicted some wounds, one of them gave him, complaining loudly, a thrust; yet immediately, as they ran away, their victim rose and pursued them, so I cannot tell whether or not he was wounded by that last blow.

An eighth curious characteristic is that on all occasions I have extricated myself when no other help seemed at hand.

And although this may be natural, the same circumstance, nevertheless, occurs so frequently and constantly, that it cannot be said or considered ordinary. Granted that to have seen that vision of the cock in dreams was natural, to have seen it so often and unremittingly in the same way may surely and justly be called a prodigy. Similarly, to throw in a fair game at hazards only three spots, when something great is at stake, or some business is the hazard, is a natural occurrence, and deserves to be so deemed; and even when they come up the same way for a second time, if the throw be repeated. If the third and fourth plays are the same, surely there is occasion for suspicion on the part of a prudent man. Likewise, in the case of a happy issue, when a favorable turn breaks like the dawn only after all hope has been abandoned, this necessarily seems to be done entirely by the will of God. To demonstrate this point I shall narrate two happenings sufficiently illustrative.

It was in the year 1542 of our salvation, and summertime, when I was wont to go day by day to the home of Antonio Vicomercati, a patrician of our city, and there to spend the whole day at a game of chess. We played, moreover, for from one to three *reals* in a single game, and to the extent that, since I was in the habit of winning, I would take away with me nearly one gold crown every day, sometimes more, occasionally less. His losses, in this manner, he accepted agreeably; for my part I found both his loss and the play itself to my liking. By this practice, I so degraded myself that for two whole years and a few months, besides, I had no concern for practicing my profession, nor for my means of livelihood, all of which were lost except that which I have just mentioned. I had no thought for my reputation nor for my studies.

On a certain day near the end of August, Vicomercati deliberately made a new resolution, either because he repented of his now almost steady losses, or because he deemed it just to me, that he would not be enticed by any reasoning or any oath or curse, and that he would even force me to swear never to return to his house, in the future, for the sake of gambling. I gave my oath, by all the gods; and that was the last day. Immediately I devoted myself to my studies.

And, look you, in the beginning of the month of October when the University at Pavia had closed its doors because of the war, and all the professors had withdrawn to Pisa, the office of teaching in my own city was offered me. This opportunity, entirely unexpected, I embraced because it did not take me away from Milan; whereas had it been necessary for me to leave Milan, or to meet a competitor, I should by no means have accepted.

Since I had never lectured except on mathematics, and then only on holidays, I would lose the stipend from the city for those days, and would be put to the inconvenience of moving my family and furniture, and would run some risk of endangering my reputation, in leaving Milan. When, for these reasons, in the following year I felt unwilling to withdraw from my city, on the night before the day on which the Senate sent a messenger to learn from me what decision I had reached, my whole house—it seems beyond belief—fell down from its very foundations, nothing more than the bed in which I, my wife, and children were sleeping, being saved from the ruin. And so what I should never have done voluntarily, nor have been able to do without some sign, circumstances forced me to accept. All who heard of the event were dumbfounded.

I shall relate another strange thing—my whole life has been filled with occurrences of this sort—but of a somewhat different nature. I had, for a long time, been having trouble, almost to the point of despairing of my life, with a kind of spurious empyema, which I have mentioned before. I had read in some notes collected by my father that if anyone at eight o'clock in the morning of the first day of April should, on bended knee, beseech the Holy Virgin that she might intercede for a legitimate petition, and if he add, as well, the *Lord's Prayer* and the *Hail, Mary*, the petitioner would obtain the favor sought. I observed the day and the hour; I went through with the prayer, and forthwith, on Corpus Christi, in the same year, I was freed from my trouble. But on another occasion long afterwards, mindful of this healing, I said a prayer for my gout—and quite properly so, for my father had added two examples of people who had been cured—and derived so much benefit that I was cured. But in this event I used the aids of my art.

Now I shall narrate four examples or singular proofs of certain peculiar circumstances surrounding the life of my older son. One happened on the day of his baptism; a second in the last year he lived; a third, in the very hour in which he confessed the crime for which he died; a fourth which began on the day he was taken prisoner and lasted until the day of his death.

He was born in the year 1534 on the fourteenth of May, and I feared for his life until the seventeenth of May. It was the Sunday on which the boy had been baptized. A brilliant sun was shining in the bedroom between eleven and twelve o'clock. All there, as was the custom, stood around the bed where the mother lay, with the exception of the houseboy. The linen curtain was drawn back from the window, and was clinging to the wall. A huge wasp flew into the room, and went winging around the baby. All who were there feared it, yet it harmed no one. Soon it became entangled in the curtain, where it buzzed with such a booming sound that you would have declared a drum was being beaten. We all ran thither, but found nothing. It could not have gone outside since we were all watching it so closely. This incident filled us all with premonition that something dire would eventually happen to the boy, but we imagined nothing so bitter as actually befell.

In the year in which he died I had given him a new cloak of silk, the customary garb of the physician. Again it was Sunday; he went to the Porta Tosa where there was a butcher, and, as usual some pigs before the gate. One of these struggled up out of the mud, and so alarmed, soiled, and attacked my son —since it took not only my son's servant, but the butchers and neighbors to drive off the animal with staves—that the incident appeared to be a veritable portent. The pig, at length, wearied of the struggle and of pursuing my son as he fled, ceased. From this accident the boy came to me sadder than usual, and told me all about it, questioning what it might presage for him. I replied that he should have a care lest by leading the life of swine he might come to the same end. Yet except for love of gambling and of the delights of feasting he was an excellent young man and led a life beyond reproach.

Again, it was in the month of February, and the beginning of the following year. I was living and lecturing at Pavia, when by the merest chance, looking closely at my hands, I observed on the ring finger of the right one, at the root, the image of a bloody sword. Suddenly I felt sore afraid. What now?

That evening a messenger appeared, afoot, bearing letters from my son-in-law, advising me that my son had been arrested and that I should come to Milan. I went the following day.

For fifty-three days that sign gradually ascended my finger, extending from the base; and, lo, on the last day it reached the tip of my finger where it blazed a blood-red flame. I, who did not then anticipate what actually came to pass, terrified and beside myself, knew not what to do or say or think. At midnight my son was beheaded, and in the morning the portent had almost disappeared; on the second day following it had vanished.

Before this, also, during the twenty days, more or less, in which my son had been detained in prison, while I was studying in a library, I heard a voice as of one hearing the confession of pitiable men soon to perish. At once my very heart was laid bare, torn and rent and fury-driven. I leaped from the spot into the court-yard where others of the family of the Palavicini, in whose home I was staying, were seated. Not unaware of how much I might intercede in the case of my son if he had not pled guilty to the crime, and if he were innocent of it, I cried out, "Alas, because he is party to his wife's death, and has now confessed, he will be condemned on a capital charge and beheaded."

Thereupon I took my cloak from the servant and set forth to the market-place. Half-way there I encountered my son-in-law, whose countenance was most sad. He asked me where I was going. I replied, "I have a premonition that my son, having knowledge of the deed, has confessed all." He replied, "It is so. He has but now confessed." A messenger whom I had sent on ahead then came running up and recounted all that led up to the act.

Among the qualities peculiar to me, according to my nature was this: my flesh gave off, somewhat, an odor of sulphur, incense, and other chemicals. This happened, for the most part, about my thirtieth year when I suffered with a serious illness. When I had been restored to health my arms seemed to smell strongly of sulphur. At that time also, I was afflicted with an itching of the skin, but this symptom has disappeared with the approach of age.

Another extraordinary thing about me was that when I, free from all cares and with the help of masters, used to study Archimedes or Ptolemy, I understood neither of them; now, in the fullness of my days, although I have laid aside these works for perhaps thirty years, and although I am claimed by much business, beset by cares and without assistance, I follow either of them with full understanding.

DREAMS

Does there not seem to be that about dreams—the fact that they have been so real—which can be worthy of admiration? I have no desire to dwell upon the insignificent features of dreams (what would be the use?) but rather on those important aspects of dreams which seem to be the most vivid and determining. Here is an example: about the year 1534, when I had not yet come to any decision about my life, and all things were going daily from bad to worse, in the gray of one morning, in a dream, I saw myself running along at the base of a mountain which rose on the right hand. Around me surged a thronging multitude of every station, sex, and age—women, men in their prime, old men, children, infants, poor and rich, in divers modes of raiment. I asked, thereupon, whither we were all running, and one of the throng replied: "To Death." I was dismayed by this, and since there seemed now to be a mountain on the left, I whirled about, so that I should have it at the right in order to grasp the vines which, on the midslope of the mountain, and at the very spot where I stood, were covered with withered leaves and entirely stripped of grapes, even as we are accustomed to see them in the autumn. I began to climb. At first this was arduous on account of the acclivity or slope of the hill which rose sheerly from the base; soon, however, having mounted above this, I climbed upward through the hills with ease. When I stood at length on the summit, I was about to pass, as it seemed, beyond the dictates of my volition; sheer, naked rock rose around me, and I seemed on the verge of plunging headlong into the depths of a hideous abyss and a dismal chasm. Forty years have passed away, but even yet the memory of that dream fills me with gloom and terror.

And so I turned away to the right where a lonely plain stretched in view covered with heath. Thither, urged by my fear, I hastened, with never a thought of what the way might lead to. I perceived that I was near the entrance of a cottage belonging to some country estate, and thatched with straw, reeds, and rushes. I was holding a boy by the hand who, dressed in an ash colored suit, seemed to be about twelve years of age. At that point, I was aroused from my sleep and dream in the same instant.

From this vision I read a manifest prophecy, pointing to the immortality of my name, to my arduous and never-ending labors, to my imprisonment, to the overwhelming fear and sadness of my life. In the flinty rocks I saw a forecast of my hard lot, and in the dearth of productive trees and useful vegetation, the emblem of my barren existence, which eventually became happy, smooth, and easy. The dream pointed, also, to my lasting glory in the future, for year by year the vine renders its harvest. The boy, if he were indeed a good spirit, was an auspicious omen, for I held him closely. If he represented my grandson, the omen was less favorable. The little house in the solitary place was symbolic of some hope of tranquillity. But that overpowering horror at the edge of the abyss might have foreshadowed the ruin of my son—it is not becoming to believe that he would have been omitted in the omen—who married and came to a disastrous end.

This dream I had, as it happened, at Milan.

I dreamed a second dream not long after while sojourning in the same place. It seemed to me that my naked soul was in the Heaven of the Moon, liberated from my body and solitary, when, as I seemed to be lamenting on account of my state, I heard the voice of my father speaking to me: "I have been appointed guardian to you by God; all these spaces are filled with spirits, but you see them not, even as I am invisible to you, nor is it lawful for you to address them. You will remain in this Heaven seven thousand years, and the same number in star after star until you reach the eighth; after the eighth you will come into the Kingdom of God."

Of this dream I have made the following interpretation: the soul of my father is my tutelary spirit; what more loving or more gracious interpretation could there be? The Moon signified grammar; Mercury, geometry and arithmetic; Venus, music, the art of divination, and poesy; the Sun, the moral life, and Jupiter, the natural. Mars represented medicine, and Saturn, agriculture, knowledge of plants and of the remaining simple arts. The eighth orb stood for the final harvest of all understanding, for natural science and various studies. And after these things I shall rest serenely with my Prince, the Lord.

This dream was, as it were, set forth—even though I did not notice it at the time—in the seven divisions of the *Problemata*, for which the time of completion for publication was at hand.

I seemed, also, in my dreams at times to recognize a youth who addressed me with flattering words, and whom I could not remember in the least when I was awake. When I asked him who he might be and whence he had come, he reluctantly responded: "Stephanus Da Mes." This name has no significance as a Latin word beyond the fact that it has a foreign sound. I used to ponder thus: *Στέφανος* signifies *crown* and *μέσος* the *middle* or the midst.

Another occurrence foreshadowed the life I was about to lead Rome. In the year 1558 on the seventh of January while I was living at Milan without any duties beyond my own private affairs, I seemed in a dream, to be in a certain city filled and adorned with many palaces. And, among other wonders, I spied a dwelling very like to a gilded house which I actually came upon sometime later when I was at Rome. It seemed to be, moreover, a feast day, and I was alone with my servant and mule. Both were well out of sight, around the corner of a house; yet I seemed to hear the voice of the servant from afar. Since in that locality but few passed in the street, I, full of curiosity, was trying to inquire from all who went along what the name of the city might be. Nobody could inform me until a certain old woman said it was called *Bacchetta*, in Latin the *virga* or rod with which boys are usually punished; long ago it was called the ferule; Juvenal says:

Et nos ergo manum ferulæ subduximus.[1]

And so I went anxiously about seeking someone who would tell me the true name of the city, for I reasoned thus with myself: "It is not this barbarous name; I have never known that there was such a city in Italy." These remarks I addressed to the old crone. She went on, "In this city are five palaces"; to which I replied, "Yet I have seen more than twenty." Thereupon she insisted, "But there are only five." And with that, trying vainly to get to my servant and mule, I awoke. I have no certain theory which I may bring forth, about the meaning of this dream; only this was made evident: Rome was plainly the city of the dream, and I knew what that word Bacchetta meant. One man said it was a Neapolitan word. This vision came to pass under auspices or for reasons somewhat confused, or else by divine intention.

I had a warning, also, in 1547, in the summer at Pavia while my younger son was sick, lying, as it were, at the point of death, that I should be bereft of the object of my affection. I was awakened, and straightway the nurse came running to me, saying: "Rise, sir, for I think Aldo is going to die!"

"What is the matter?"

"His eyes are rolled back and he does not make a sound."

I rose immediately, administered a powder of pearls and gems in which I had faith. He vomited. Again I gave the powder, and this time he retained the dose, slept and perspired, and within three days he was well.

It is not unmeet that men of conscientious conduct, firm in faith in God, discreet in judgment and in looking toward salutary ends should have such experiences: provident heads of families, as it were, taking advantage of every opportunity for protecting, by a course especially consistent with medical science, the body and that vital principle mingled therewith. Others vainly attempting an imitation of such a course make of themselves a laughing stock, like empty prophets; for things of that kind cannot be reduced to an art. Here, however, it was a case of that undutiful son of mine who brought me so much trouble.

Many amazing dreams, besides these, have I dreamed, strange beyond belief, which I do not wish to relate.

These extraordinary experiences, then, have been mine: intellectual powers of differing character, and dreams, as well as four intuitive flashes or premonitions, three of which I have already set forth: one resulted in my sudden withdrawal from the contract I had made at Magenta; another came before the storm I experienced on the lake; a third was when my house fell to ruins. The fourth is the remarkable incident of my garter, concerning which I have written in another place. Therefore must I recognize these experiences as special dispensations from the Lord. Nor has anyone shown any inclination to ascribe whatever has happened to me of this sort, whether by chance, by dreams or by premonition, to any natural cause; if anyone should take such an attitude, he would be exceedingly, and even, perchance, gravely in error. Neither will anyone, much less, wish to ascribe these things to my own merit. They are rather gifts of a bountiful God, who is in debt to no one, much less to me. Those are even greatly mistaken who attribute my experiences to my own diligence or efforts or zeal, which would scarcely have been able to effect a thousandth part. And farthest from right are those who have fancied that I have fabricated these things out of an empty ambition for glory; I shrink from the thought of it. Finally, why should I wish to mar by trifling singsong and false representations that virtue not only natural to me, but which I know derives from God?

38.

FIVE UNIQUE CHARACTERISTICS BY WHICH I AM HELPED

UP TO THIS point I have discussed myself as an ordinary human being, so to speak, and as a man even somewhat lacking, in comparison to other men, in natural endowment and education. Now it is my purpose to give an account of a certain extraordinary characteristic of my own, one indeed which is the more unusual for the reason that I know it to be part of me, yet the essence of it I am not able to define. And it is *myself*, although I am not aware that such virtue has originated from myself. This power is present when needed, but not in evidence when I so will. What comes to pass as a result of it is something more than my natural powers could effect. This endowment was first observed toward the close of the year 1526 or at the beginning of the following year, so that more than forty-nine years have elapsed since I have been aware of it.

I am conscious that some influence from without seems to bring a murmuring sound to my ear from precisely that direction or region where some one is discussing me. If this discussion be fair, the sound seems to come to rest on the right side; or, if perchance it approaches from the left, it penetrates to the right and becomes a steady hum. If, however, the talk be contentious, strangely conflicting sounds are heard; when evil is spoken, the noise rests in the left ear, and comes from the quarter exactly whence the voices of my detractors are making disturbance, and, accordingly, may approach from any side of my head. Frequently, when the discussion comes to an evil conclusion, the vibration in my left ear is resumed in the very moment when it seems properly to be ended, and the sounds are multiplied. Very often when the discussion about me has taken place in the same city, it has happened that the vibration has scarcely ceased before a messenger has appeared who addresses me in the name of my detractors. But if the conversation has taken place in another state and the messenger should appear, one has but to compute the space of time which had elapsed between the discussion and the beginning of the messenger's journey, and the moment I heard the *voices* and the time of the discussion itself will fall out the same. One may see, besides, that the decision was carried out in the manner which I had inferred from the character of the noise in my ear. This manifestation lasted up to the year 1568 until about the period immediately followed by that conspiracy,[1] and I was surprised that it should cease.

A few years later, eight perhaps, that is, about 1534, I began to see in my dreams the events shortly to come to pass. If these events were due to happen on the day following the dream, I used to have clear and defined visions of them just after sunrise, so that even on one occasion I saw the motion for my admission to the College of Physicians straightway brought to vote, to a decision, and the motion lost. I dreamed, as well, that I was about to obtain my appointment to the professorship at Bologna. This manifestation by dreams ceased in the year just preceding the cessation of the former manifestation, that is, about 1567, which was the year Paolo, one of my wards, went away. And so it had lasted about thirty-three years.

A third peculiarity is an intuitive flash of direct knowledge. This I employed with gradually increasing advantage. It originated about the year 1529; its effectiveness was increased but it could never be rendered infallible, except toward the close of 1573. For a period between the end of August of that year and the beginning of September 1574, and particularly, as it seems to me, now in this year 1575, I have considered it infallible. It is, moreover, a gift which has not deserted me, and it replaces the powers of those two latter faculties which did; it prepares me to meet my adversaries, and for any pressing necessity. Its component parts are an ingeniously exercised employment of the intuitive faculty, and an accompanying lucidity of understanding. An altogether pleasing faculty, it is far more profitable for my influence, my training, for gain, and for confirming the results of my studies than the other two special faculties together. It does not divert a man from his pursuits or from intercourse with his fellows; it renders him ready for every event; as an aid to the composition of books it is invaluable. This faculty seems, as it were, the most elemental quality of my nature, for it exhibits at one and the same time the essence of all the qualities which compose my nature; and if it be not a divine endowment, it is certainly the most highly perfected faculty which man may cultivate.

Meanwhile there was a fourth peculiarity which began to function in the year 1522 and which lasted through to the year 1570 or 1573. This was plainly enough given me, I believe, as a certain assurance; for after my life was saved, when I had seemed past all hope, it came to confirm my faith and strengthen the conviction that I am God's and that he is all in all to me, and to keep me from committing anything unworthy of so many benefits. And if anyone should ask why all men, or at least a few do not have these or similar experiences, my reply is, "How do we know but what many are aware of similar manifestations, or have spiritual experiences clear to themselves, although not so to us?"

Again, "But what indication of loving favor toward you is that cruel death your son suffered?" "If anyone is able by any other way, without suffering, to gain immortal life, I claim that way; if not, what special dispensation should I receive? For death is always bitter, and equally bitter is that daily and inevitable expectation of the end which is more than death's counterpart, and almost death itself."

A fifth peculiar feature of my life, and one which has never ceased to attend me is this: whenever my personal affairs have been in a state nothing short of desperate, I have been swept up on a wave of fortune; these affairs had no sooner become fair and flourishing than I sank into deep waters. Even as galleys, tempest tossed, are driven now from the depths to the crest of the surge, now plunged from on high into the yawning gulf, so has it gone with me throughout my career. O, how often have I lamented my so wretched condition, not merely because everything went to ruin, and every hope of safety would be swept away, but because not even when I sought by thoughtful calculation to put my affairs in order, would I ever find a way out; then, with no effort or labors on my part, within two or three months I saw that all was changed, so that I believed there were greater influences than my own will in the matter, and that whatever came to pass would be the result of an outer force; and this happened so many times that I am embarrassed to number them. By this vicissitude of fortune, also, it often happened that all things collapsed at the same time.

39.

ERUDITION, OR THE APPEARANCE OF IT

THE PURPOSE of this chapter is to discover whether I actually know anything, or whether I only seem to know. Grammar I have never learned; likewise I have never learned the Greek, French, or Spanish languages, but the ability to employ these has been given me; how, I cannot explain. I did not, furthermore, gain any acquaintance with rhetoric, nor indeed optics, nor the theory of equilibrium, because I did not expend any amount of study on these subjects. Nor did I give any attention to astronomy because it seemed too difficult. On the other hand, though I had but meager talent as a musician, I was unequaled in the study of the theory. In geography, in philosophy based on controversy, in the doctrine of morals, in jurisprudence and in theology I have not exerted myself; for I considered their matter too comprehensive, not in accord with my purpose, and of such a nature as to claim a man's undivided interest.

Much less did I devote myself to any doctrine, evil, pernicious, or vain, such as chiromancy, the science of compounding poisons, or alchemy. To physiognomy, likewise, I gave no attention; for it is a long course and most difficult, one which requires exceptional powers of memory and very ready perception, which I scarcely believe are my endowment. Nor did I study the art of magic which deals with enchantments, nor yet have aught to do with summoning the spirits or the ghosts of the dead.

Even to those subjects of more praiseworthy character I have given very little study. Faulty memory kept me from identifying the plants; I did not concern myself with agriculture, for in that pursuit it is more necessary to practice than to know the theory. From anatomy many considerations frightened me away.[1]

I was not given to composing songs except when it could not be easily avoided, and, at that, I composed very little.

Why, therefore, is it that so many have attributed to me interests on which I do not spend thought, unless it be that they may diminish my reputation in the art of medicine?

Pluribus intentus minor est ad singula sensus.[2]

That branch of astrology which teaches the revealing of the future I studied diligently, and much more, indeed, than I should; and I also trusted in it to my own hurt. The part of astrology which is more of a natural science was not a subject in which I had practice, for it is but three years, that is, at about my seventy-first year, since I acquired any knowledge of it.

I was versed, then, in geometry, in arithmetic, and in medicine—both in theory and in practice. In logic I had even more skill. Natural phenomena, as concerns the properties of substances and things of their same nature, I studied; as, for instance, the fact that amber gives proof that heat was stored within it, and why; also, if it may be included, the technique of the game of chess. To this list I may add a practical knowledge of the Latin tongue and of certain others, and finally, the theory of music.

Navigation, however, I never touched upon. As to military science, there is no reason why I should include it among the arts; and besides, owing to the great number of difficulties it presented, I was not in the least acquainted with it, as was the case also with architecture.

There are also certain pseudo sciences, as the practice of symbolic writing, composition, and interpretation.

In my own field I am destitute of any practice in surgery.

If then you place the number of important branches of learning at thirty-six, from twenty-six of the studies and from any acquaintance with them I have refrained. To ten I devoted myself.

Some have, moreover, thought my understanding and knowledge greater because of an outer expression of it. The latter is sustained by serious and lasting contemplation; by inferences from many facts well known; by the choice of better principles and not by a zeal for controversy, as in the case of Galen. It rests neither in notions too general in application, nor on assumptions in any degree false—though my devotion to truth permits me to digress in some points a trifle, not much, as some would have it, from the prevailing opinion—nor in imaginary assumptions, as with Plotinus. This outer expression consists, rather, in a nicety and stability of judgment, in experience of age, in foreknowledge and in the exercise of those five means I have already so often said I use as aids.

Let me add, indeed, to these ten branches of learning which I have mastered, the knowledge of many historical facts; though this information belongs properly to no particular study, it adds much, nevertheless, to the embellishment and adornment of the facts treated in these studies.

These suggestions I have wished to include for exhorting, as it were, each one that he rather—inasmuch as our life is brief and filled with obstacles and disadvantages—be willing to devote his energies to a few things than to many, and to these diligently and assiduously; that he should choose above all, pursuits useful to his fellowmen and, primarily, to himself; that he should accept inevitable inferences and true beginnings; nor in provocation, as it were, or for the sake of popularity, cast aside established good, but make trial of now this, now that, as one or the other means is better.

If a care for fame leads you on, or if you dare to hope for

some advantage from this, it is better to polish to a perfect finish one invention of the mind than to pursue a thousand purposes and overtake nothing.

> Sæpius in libro memoratur Persius uno:
> Quam levis in tota Marsus Amazonide.[3]

In this respect we see Horace especially successful, owing to one composition, not a long one, to be sure, but carefully perfected and pure gold. And now, that the poet may make good his boast: *Dum Capitolium scandet tacita cum virgine pontifex*,[4] that he would prevail over time by his writings, nor feel the Stygian wave's compelling power, it has surely come to pass that though the *Pontifices* have ceased to make the ascent to the Capitolium, the fame of Horace still is flourishing. Painstakingly, then, I have advanced arithmetic, as a science, tenfold and medicine not a little.

It is the part of a serious man to make haste to accomplish what he would. And to this end a vast amount of reading is necessary; when I am devouring a mighty volume in three days of steady reading, some suggestion as to the contents is necessary, by indicating for omission parts very trite or of little use, or by placing a dagger to mark obscure passages which may then await an occasion for investigation.

As to my arrangement in composition, it is my custom to preserve the same style for the end of a sentence as for the beginning of the next. I have the most inspired authors as authority for consistency of this sort. Let the writing be, also, polished, pure, coherent, unified, in the Latin tongue, and of a style guarded by propriety of diction; let the thread of the composition and of the sense be drawn from one source. Those arts which are, to be sure, not finite, as geometry and arithmetic, do not suffer adornment; others, contrarily, are rather subject to division and embellishment, such as astronomy and jurisprudence.

40.

SUCCESSES IN MY PRACTICE

1. IN THE summer of the year 1537 or 1538 a close friendship existed between me and Donato Lanza to whom I had been able to give relief for a spitting of blood with which he had been afflicted for many years. As he enjoyed the confidence of Senator Sfondrato, a private counsellor of the Emperor, he had often urged this nobleman to use my skill toward the curing of his elder son, a child suffering with puerile convulsions, and to be counted rather in the number of the dead than the living. He was imbecile, mentally defective, and deformed as well, yet in spite of this weakness the child did eventually get well.

Sfondrato's other son, the younger, a baby in the ninth or tenth month, was taken with fever. Luca della Croce was treating him, and holding out hope for the child's recovery, as is the custom. Della Croce was Sfondrato's intimate friend, inasmuch as he was a procurator of the College of Physicians of which Sfondrato was a patron, and the two were bound by ties of long association, and interchanges of services and benefits. Suddenly very sharp convulsions, with continued high fever, seized the baby. Since with these symptoms the danger of death was plainly imminent, della Croce urged that Ambrogio Cavenaga be summoned; while Sfondrato, mindful of Lanza's recommendation, suggested me. We went into consultation at eight o'clock in the morning, together with the father of the child. Thereupon della Croce briefly set forth the history of the case, because he knew Sfondrato to be a man of intelligence, and because he was himself sincere and learned. Cavenaga offered, meanwhile, no opinion, it being his place to speak last.

Thereupon I said, "You observe that the child is suffering with opisthotonos." At this term the senior physician looked startled, as if I had intended to confound him by the use of an obscure expression. But della Croce at once carried off the embarrassment by saying he understood that to be a nervous spasm in which the body is bent backward. I replied, "Precisely, and I shall now show what I mean."

Thereat I lifted the baby's head which hung backward, a state which the doctors and others thought was due to an imbecility which caused the head to droop with its own weight. I ordered that the head be elevated to its former position, gently and gradually, however, for any other method would have been of no avail. All present looked on in admiration, and more particularly the father. Then della Croce suddenly uttered these words: "Horbon! Look you!" in the phrase of a man rousing someone, or of a person beginning an address. "In the diagnosing of diseases Don Girolamo has not an equal!"

Sfondrato quickly spoke then, addressing me. "Since now you have recognized the disease can not something be effected by medication?"

As the others were silent, lest I should impair the reputation I had already gained by seeming to promise more than I could well carry to a successful end, I turned to my associates and said, "You know what Hippocrates says in regard to this: *Febrem convulsioni.*"[1] And I cited the aphorism.

Thereupon della Croce—who, with his characteristic sense of fitness was desiring to maintain a friendly spirit, if possible, so that, should the boy be cured, his aforetime favor might be maintained, or in case the child died, that he might not seem to have been envious of the reputation of a rival—

assigned the case to me. The senior physician, Cavenaga, acquiesced, for they knew there was more to be anticipated from discreet praise than from disagreement.

Thereupon I prescribed a fomentation of gauze moistened in linseed-oil, and in oil of water-lilies, to be applied, and gave directions that the baby be gently treated until his neck resumed its natural position; that the nurse should eat no meat and the child be given no food nor drink other than her milk, and not too much of that. I told them to put him in his cradle, keep him warm, and rock him easily until sleep should creep over him.

When the other doctors had withdrawn I recollect how the parent of the boy said to me: "I give this child to you for your own son."

To which I replied, "You are not well mindful of his future in wishing to substitute a poor father for a rich one."

"But I know," he went on, "that you have treated him as if he were your own, never fearing that in so doing you might give offense to those." He meant the physicians.

"On the contrary," said I, "I am pleased to confer with them, and would be glad to have them for associates in all cases, and gain their support."

And I so combined my phrases as to give him to understand that while I did not despair altogether of the cure of his child, I was not overconfident; that I was more discreet than profoundly learned or able to be of any great service by way of knowledge gained from experience with the disease. The cure was successful, for as it happened the fever had been running for fourteen days, and the weather was warm; within four days the baby began to improve.

It was in view of this—not so much because I diagnosed the disease, I conjecture, for my own particular experience might have helped therein; nor because his son was restored, for these things can be attributed to chance, but because the boy recovered in four days, whereas the doctors had tormented his brother for more than six months and in the end left him half dead—in view of this, it was, that he looked upon me with admiration and thereafter preferred me to all others. It may be assumed, at least, that he had great respect for me because della Croce maintained an envious and unfriendly spirit toward me throughout his term as procurator of the College of Physicians; and he declared to Cavenaga, in the presence of my patient's father that he could in honor say nothing more than he felt obliged to in praise of a man who did not enjoy amicable relations with the College. It was plain that jealousy and bitter rivalry and not the conditions of my birth stood in the way of my admission to the college. Sfondrato was so impressed with the case of his son that he told the whole story in the Senate, and used his persuasions upon the Governor of the State and other ministers and men in the purple robes of rank to such effect that a way into the College of Physicians, from which I had been rejected by so many votes, and finally on so many charges and excuses, was opened. He brought it about that I might be admitted to the post of lecturing publicly in the Academy, and even be recompensed with a gratuity, and be received by all.

2. My next case was that of Archbishop Hamilton of Scotland, who, then in his forty-second year, had been for ten years suffering with asthma. Having tried first the physicians of the King of France, and after that the physicians of the Emperor Charles V, but to no effect, he sent for me at Milan, asking that I proceed to Lyons. Once there he sent an additional three hundred pounds, requesting me to continue on my journey as far as Paris, and even, should he be prevented by the wars from meeting me there, on to Scotland.

I went. The Archbishop's doctor was pursuing a course of treatment in accordance with the decisions of the Council of Parisian Physicians. That the patient did not mend gave cause for reproach, and I was finally obliged to express myself as to the reasons for his continued indisposition. Whereupon the Bishop vented his indignation upon his physician, and the latter upon me, because I had uncovered the trouble; and so I was in apprehension from the latter, and accused of procrastination by the former, the more so because when I had begun my treatment, the patient had improved. In the midst of such disturbing circumstances, I asked leave to withdraw, a favor which was not very graciously granted. On leaving, I prescribed a regimen whereby he regained his health after two years. I tarried with him seventy-five days; and after he saw his expectation of being cured was in the process of realization, Michael, his lord chamberlain, was commissioned to offer me generous inducements to return as private physician to his household, but I did not accept. The Archbishop paid out for this attendance eighteen hundred crowns of gold, of which fourteen hundred came to me.

3. In my own country, I cured of a cutaneous affection of two years' standing, Francesco Gaddi, Prior of the Monks of St. Augustine. Within six months, he was sound; but this man and my preceding patient—Oh, wretched fate of humankind!—were freed from disease to perish ten years later, by violent death, as a result of political strife.

4. Martha Motta was healed by my treatments in two years, although for thirteen years she had been confined to a chair and had never taken a step. Although in two similar cases, within about ten years, an attempt to cure had resulted in violent death, this woman, when I left Milan, had already lived twenty-three years; and, though in truth with a slight stoop, had been able to walk where she would throughout all this time.

5. Giulio Gati I cured of consumption, and not long afterwards he became the childhood tutor of the Lord of Mantua.

6. The son of Giammaria Astolfo I relieved of chronic fevers.

7. Adriano, a Belgian, I cured of an empyema; and thereafter he was grateful, devoted, and full of services toward me, to such an extent that I should like to have the same experience with an Italian.

8. Again, Giampaolo Negroli, a merchant well known throughout the whole city, having sought a cure from all the leading physicians, was at length given up as hopelessly consumptive. My treatment of his case secured for me his firm friendship.

9. Gasparo Rolla, a shop-keeper, who had already been for the period of a year but little better than a millstone, and plainly never able to make a move through his own efforts, I cured. He remained, however, twisted and wry-necked.

10. What of the fact that I was not wont to lose a single fever patient, and of those sick of other ills, scarcely one in three hundred? Proof of this are the statements accompanying death certificates as preserved in the Public Health Records. This custom of the public officials is well known to every citizen. However, although this proof is not at hand, still I think it neither proper nor expedient to bring forward testimonials from other sources, nor to boast, especially since physicians take little stock in these methods.

11. I was also summoned from my professorial duties in the University of Pavia to attend professionally the Duke of Sessa.[2] I accepted as a fee one hundred gold crowns and a gift of silk.

12. I was similarly called from Bologna to Modena to Cardinal Morone, from whom I was loath to take a fee; for I

was aware that I owed more to his patronage than he to my services. In these two illustrious men I found firm supporters and ready helpers, and I was sustained by their personal efforts in my behalf.

Altogether I restored to health more than one hundred men given up as hopeless, at Milan, Bologna, and Rome. Nor should it seem a marvel for me to have shown myself so lucky and successful in the practice of therapeutics, since I claim for diagnostics a very special place in the field of medicine.

Two offers which I made publicly at Bologna are evidence of my success; I said that I would cure every sick person who should come in time into my hands, who was not more than seventy years old, and not younger than seven (in my *Prognosticorum*, where I frequently mention this, *seven* should be read in place of *five*); nor anyone with a disease pendent on procatarctic causes such as a wound, a blow, an accident, a fright, or poison; anyone (I emphasize this especially) in possession of his faculties, and not an invalid such as a consumptive, or with a scirrhous liver or a deep ulcer in a dangerous place; or if with a very large stone in the bladder, or with epilepsy.

The other offer—in which, however, I was free to take up the hazard or not—was that if anyone fell sick unto death, I would show where the seat of the disease was, and if, after the patient's death, it was found that I was in error, I was to be held for a penalty of one hundred times the moneys taken in stake. Accordingly, very many, openly eager at first to be able to prove that I had been mistaken, had dissected bodies, as that of Senator Orsi, of Doctor Pellegrini, and of Giorgio Ghisleri. In the last case, does not my prediction that the source of the ailment would be in the liver seem astonishing, when the urine was in no way affected? That the abdomen, which had pained constantly, would be unimpaired? After that, though they secretly examined many other cases, they never found me in error, nor did they dare to accept my challenge, nor recommend that it be accepted.

13. But to return to my successful cures: At Bologna I restored to health Vincenzo Torrone who had been suffering for the space of a year with an affliction of the hip. He had been confined constantly to his bed, a fact which had contributed in no way toward a cure, and had not even lessened his pain.

14. From a similar disease I cured, in the dead of winter, the wife of Claudio, a merchant of the city.

15, 16. Just so, at Rome, I attended professionally a noble lady, Clementina Massa, and also Giovanni Cesare Buontempo, a jurisconsult. Both had been sick for almost two years and were reduced to the most distressing circumstances. They had tried all the more renowned physicians of the city, but I treated them so effectually that they are living yet today.

In attending, furthermore, a community of Spanish people living at Milan I gave proof of some extraordinary facts: first, what of this, that no one was able to boast that he could heal anyone whom I had given up as incurable, although I restored a great number of patients given up as hopeless? And, note, as I have elsewhere declared, that *chance* has no place in art. Wherein should *Fortuna* attend the barber in order that he may shave, or a musician so that he may sing and strike the chords? Similarly in medicine *chance* is not to be counted on, yet there are three respects in which medicine, setting aside any considerations of skill, seems to be subject to chance: first, the fact that all points of the medical art are not so simple and clear as in the art of a shoemaker or a barber. If any physician, then, has come, in his practice, upon a rather grave case, or met with a disease which is very complex or deep seated, and has not been able to offer much in the way of help, it is due not to the shortcomings of medical art, but to the inexperience of the one engaged in practicing the art. Medicine is, moreover, a many-sided art, to such an extent, indeed, that both at present and in the past it has been divided into the branches represented by surgeons, oculists, physicians, the druggists, those skilled in the knowledge and use of herbs, and the bone specialists. Each of these special fields is shared by many groups devoted to some particular phase of their specialty. Accordingly, if any doctor's practice brings him in contact with a disease with which he has long been familiar, and has successfully treated, he can be called lucky; otherwise, it is just the contrary. The doctor, furthermore, must needs deal with medicaments, nurses, assistants, pharmacists, surgeons, and those who prepare the food, and see that all externals proceed without error. He must have a care for heat, water, the bedroom, sanitation, silence, and the patient's friends. Fear, depression, or a fit of anger may bring about a patient's death even though the disease of which the patient is suffering is curable.

Yet withal, in short, since it is an art, medicine is not a subject of mere chance. In so far, however, as any art is many-sided, as medicine is, or results from many agencies working, together (as soldiering—which is, to be sure, of a single character, albeit of many agents of the same type) or is the product of the work of one agent (as the craft of the artificer of nails, or better, of bronzes, or as the labor of the plowman and the sower), that art is, accordingly, subjected to various vicissitudes. And the fact, too, that a doctor practices—as I have—in many localities may have some bearing on the element of chance in the art. Hippocrates avers that it has much to do with the matter. I practiced first at Venice, after that in the region of Padua at the village, or *commune,* to use the Italian term, of Sacco; then at Milan, at Gallarate, at Pavia—though here but little—at Bologna and at Rome. Abroad I practiced somewhat in France (Lyons), England, and Scotland. I keep up my practice even at this period of my life, and I am entering my seventy-fifth year, although Galen did not pass beyond his sixty-seventh, and Avicenna, that is Hossein, his fifty-seventh, and both spent their whole lives in travel. At least Galen traveled for twenty years, and Hossein always went from place to place. Aëtius[3] traveled throughout his Episcopate, Oribasius[4] was banished to Pontus, and Paul of Ægina was a wanderer.

17. But to continue, let us relate the case of Giulio Ringhieri, a young fellow-townsman. This man lived near the Church of San Giacomo on the street called San Donato in Bologna. On June 20th, 1567, when he had been lying for more than forty days with a very high fever accompanied by acute inflammation and delirium, I cured him, since with the customary formalities his case had been turned over to me.

18. What need is there to mention Annibale Ariosto, a rich and aristocratic youth? He was troubled with an abscess in the chest which had become chronic, and because he had lost flesh, had had an accession of hectic fever, had discharged as much as two pounds of matter in a single day, and was suffering with insomnia, the doctors abandoned him as an incurable consumptive, declaring his lungs hopelessly diseased. They isolated him lest the four young children of Michele Angeli, judge of Turin, should be touched by the contagion. This patient I restored to health in thirty days, and, what is more, he so put on flesh and gained in healthy color that the whole city was amazed. Nothing like these two miraculous cures was seen for fifty years at Bologna.

19, 20. Likewise there were two young men living in adjoining houses near the gate which leads to Modena—Leonardo and Giambattista—who, sick with fever and dysentery yet not troubled with a cough or any difficulty of breathing, had lingered for eleven days, though they had been given up for as good as dead. I knew they were suffering from a congestion of the lungs, and asserted that *both* could be cured by so simple a means as a discharge of mucus from the lungs, a statement which amazed the doctors. Even as I had said, *both*, within twenty-five days from the start of the trouble, after they had discharged something like a pound of mucus, recovered and were completely well within four or five days.

21. Not far from these two young men lived another, one Marco Antonio Felicino, who, as the result of an extended illness, was not able to speak. He was, moreover, the grandnephew of a senator of high rank, by reason of his very ancient family, as I learned. From a siege of uninterrupted fevers, during which he was unconscious, he escaped voiceless and weakened unto death. His case was given over by other physicians, since they acknowledged that they neither recognized the disease nor had seen its like. There were those who said the youth was reduced to such a pass from a draught of poison given him by his mistress. I restored him to self-control and speech four days after I had taken his case into my hands, and within another ten or twelve days I had cured him entirely. I believe he is still living.

The physicians bore a grudge against me on this score, because, in the treatment of such cases, I had made no explanation of my diagnosis nor any exposition of my method.

22. There was the cure of Agnese, wife of Claudio, a French merchant of our city for whom the foremost physicians had ceased to hold out any hope, saying she was moribund. And this they said with good reason, for certainly, of all my patients, I never saved the life of anyone with greater efforts than I put forth on her behalf, though I have seen many nearer death.

23. I remember cases which I have mentioned casually elsewhere—certain patients cured of mental unsoundness, of

epilepsy, and of loss of sight; people suffering with dropsy, with humped backs, with failure of organic functions, and with lameness, were treated by me to the improvement of their condition, such as (24) the sons of the woodworker who lived near the Tosa Gate. Why (25) go into detail about the little short of miraculous cure of Lorenzo Gaggi and of (26) the legate of the Prince of Mantua, or (27) about the really noteworthy healing of the illustrious Spaniard, Juara, and of all the Milanese Spaniards? Why need I more than mention the name of (28) Simone Lanza, of (29) De Mareschalchi, of (30) the daughter of Gianangelo Linato, of (31) Antonio Scazzoso? That case (32), finally, of the son of Martino the merchant is worthy of attention, and that (33) of the wife of the pharmacist of the Three Kings. Some (34) I relieved from inveterate complaints; (35) others from a bloody passage of urine. (36) From double quartan ague no one was dismissed from my care uncured.

(37) Although summoned tardily I saved all the children of the family of the Sirtori from poisoning, even after both parents had died. (38) For Agostino Fornari I treated a case of dropsy. (39) Why need I refer to that contention with Cavenaga and Candiano in the cure of Octaviano Mariano? However, to omit the mention of an almost unbelievable number of examples is not to argue that my success diminished. (40) There was the unfortunate case of Antonio Maroraggio; yet over against that failure, the in-no-wise deceptive cures of fever, plague, and gout. It was because of such a case as this that the physicians of Milan were wont to say that I happened to save my patients not because of the excellence of my art, but by mere good luck, inasmuch as the moribund came into their hands, while to me those came who were destined to be saved.

May you not be amazed, O Reader, nor suspect me to be lying! All these facts are exactly as I have stated, and cases of exactly this sort have occurred, even of greater moment and in larger number than I have set down. I have not the exact record, but I judge the number would reach 180 and pass beyond that. Neither are you to mistrust that I am indulging in vainglory, or falsely imagine that I desire or hope to be set above Hippocrates.

In the first place, as concerns misrepresentation—why should I indulge in such folly? Let anyone look into my statements; should he detect me in a single falsehood, all I have written would be of no account. Again, why should I be charged with seeking glory and gain? Should more serious diseases present themselves to me, I should deem it nothing short of a misfortune, for I do not of a truth feel confident that I could depend on my past success. Pliny and Plutarch tell how Cæsar, when he had for the fiftieth time contended victoriously against the enemy in open field, became less eager for battle lest he should mar and dishonor the glory gained.[5] How much more easily this might be my fate!

Moreover, how dare I hope that great honor may spring from my daily round of professional visits? Only with a prince for patron may this come to pass if indeed any hope of it remain at all.

Whatever in the physician's work has glory for its end is not only much the lesser part—but, for a truth, is naught. Hippocrates is not illustrious because he healed so many, for by his own testimony, on one occasion twenty-five out of forty-two died and only seventeen were restored to health. But he is great because of his contributions to the theory and science of medicine. I, on the contrary, am doubly inferior in this respect: in the first place because that good fortune in my practice seemed to have come from divine help and not from my own great learning; so many propitious circumstances, moreover, have accompanied my successes in this respect that I entertain no hope that the future may bring results equal to my labors. And it is seldom proper to boast that anything has happened according to plan or design—although in deliberation I showed myself as efficient as I could.

Yet the diseases which Hippocrates had to combat were of the most violent nature. Thessaly is a barren rocky land where the winds are biting, the wines are raw, the vegetables lacking in succulence, and water is scarce. The mode of life is uncouth; remedies there are none, nor anything choice, and the exercises are so strenuous as to be deadly. Had I been there I surely should never have been able to publish such extraordinary stories of success; had Hippocrates lived in peaceful times, in a clement region abounding in every comfort he might have practiced medicine without running on the rocks that beset a physician's course.

Let what I have written, then, be my boast in the matter of my professional success; but let it have been within the limits of moderation—since the Lord, as I now feel, shows me his favor in so far as I render a true account. "What," some one will say, "has he not in large measure—and not to say *merely*—given you satisfaction in your vain desire for success, even though he has deprived that flower of success of its fruit seeing that you have spent your career living in great poverty, with so many rivals, exposed to so many wrongs and afflicted with so many hardships!"

Finally, I know that I have obtained whatever sum of renown is mine not only without being suspected of false pretenses, but even without bearing a grudge against any man.

41.

CONCERNING NATURAL THOUGH RARE CIRCUMSTANCES OF MY OWN LIFE. CONCERNING THE AVENGING OF MY SON

AMONG THE extraordinary, though quite natural circumstances of my life, the first and most unusual is that I was born in this century in which the whole world became known; whereas the ancients were familiar with but little more than a third part of it.

On the one hand we explore America—I now refer to the part peculiarly designated by that name—Brazil, a great part of which was before unknown, Terra del Fuego, Patagonia, Peru, Charcas,[1] Parana,[2] Acutia,[3] Caribana,[4] Picora,[5] New Spain, Quito, of Quinira[6] the more western part, New France and regions more to the south of this toward Florida, Cortereal,[7] Estotilant,[8] and Marata.[9] Besides all these, toward the East under the Antarctic we find the Antiscians[10] somewhat like Scythians, and some Northern peoples not yet known, as well as Japan,[11] Binarchia,[12] the Amazonas, and a region which is beyond the Island of the Demons, if these be not fabled islands—all discoveries sure to give rise to great and calamitous events in order that a just distribution of them may be maintained.

The conviction grows that, as a result of these discoveries, the fine arts will be neglected and but lightly esteemed, and certainties will be exchanged for uncertainties. These things may be true sometime or other, but meanwhile we shall rejoice as in a flower-filled meadow. For what is more amazing than pyrotechnics? Or than the fiery bolts man has invented so much more destructive than the lightning of the gods?

Nor of thee, O Great Compass, will I be silent, for thou

dost guide us over boundless seas, through gloomy nights, through the wild storms seafarers dread, and through the pathless wilderness.

The fourth marvel is the invention of the typographic art, a work of man's hands, and the discovery of his wit—a rival, forsooth, of the wonders wrought by divine intelligence. What lack we yet unless it be the taking of Heaven by storm![13] Oh, the madness of men to give heed to vanity rather than the fundamental things of life! Oh, what arrogant poverty of intellectual humility not to be moved to wonder!

But to return to my theme. It was the 20th day of December in the year 1557. All things seemed to be going well with me. I did not sleep until midnight, yet when the desire was about to overcome me, my bed suddenly seemed to tremble, and with it the whole bed-chamber. I thought: It is an earthquake! At length sleep stole over me. In the morning when day had dawned I asked Simon Sosia—he is now here at Rome with me—who slept in the low truckle-bed, whether he had been aware of anything unusual. He replied, saying that he had felt the room and bed vibrating. "At what hour?" I asked. "The sixth or seventh," he said. When I betook myself to the public square, I asked any number of people whether they had felt the earthquake shock, but no one had.

On my return to the house, my servant, with a sad countenance, came hurrying up to me and announced that Giambattista had married Brandonia di Seroni, a girl he indeed loved but who was utterly without dowry or recommendation. Alas for grief, alas for tears! I found when I arrived that the marriage had indeed taken place. This, then, was the beginning of all manner of evils.

I reflected that some messenger from no earthly regions had that night willed to give me warning of the event which he knew had come to pass in the same evening; indeed when day had come, before my son went out from the house I had approached him, saying—not so much at that moment warned by the portent, but because he seemed strangely unlike himself—"My son, have a care for yourself today that you bring not some disaster upon yourself." I recall the very spot; I stood in the doorway. But I do not remember whether or not I referred to the portent.

Not many days later I again felt the bedroom vibrating; feeling with my hand I detected a palpitation of my heart, perhaps because I was lying on my left side. When I sat up the tremor and the palpitation ceased at the same time. I lay back once more, and both returned; wherefore I knew the one was connected with the other. I even remembered that when the tremor had occurred on the previous occasion my heart had palpitated so as to seem a perfectly natural thing; yet at that time I did not realize how it had been affected. I noted only that there had been a double vibration, one due to the palpitation of my heart, and another a spiritual reaction coming through the medium of the palpitation. I came to this explanation through a similar circumstance of a few years earlier. It used to be my wont in those days, if I was deserted by sleep in the small hours before the dawn, to be tossed and worried with the weight of my anxieties. Now, however, within the last few years, if I happen to be wakeful, I am not beset by cares. From this I judge that the insomnia may be merely an unwholesome habit rather than a spiritual visitation.

An incident of similar nature occurred in 1531 when a dog of the neighborhood, ordinarily a well-behaved beast, howled for hours without stopping—an extraordinary performance; ravens sat on the ridgepole, croaking with unwonted persistence; when the house-boy broke the kindling fagots, fiery sparks flew forth, and before I rightly knew what I was about, I found myself married! From that moment all sorts of adversities were my constant companions. Yet one must not jump at the conclusion that all these omens were supernatural. Indeed, when I was all but thirteen years old, a raven flew at me in the Piazza de Sant 'Ambrogio and tore a strip from my garment; nor did the creature want to let me go even though I pushed and pulled at him violently. Yet for all this, nothing dire befell either me or any of mine for many years.

Many a great sight have I seen, yet natural phenomena all: for one, when I was a lad I saw a star like to Venus shining so high in heaven at the twenty-second hour of the day that it could be seen from our whole city. Thereafter, I saw three suns in the year 1531 and all casting their broad beams across the eastern sky. This happened in the month of April, at Venice where, by chance, I then was. The spectacle lasted about three hours altogether.

Now it happened that some time ago, about the year 1512 in the territory around Bergamo near the Adda River, more than a thousand stones are said to have fallen in one night, following the appearance, in the early evening, of a mighty fiery pennant which was borne upon the air in the likeness of a beam of amazing proportions. One of these stones I saw when I was a boy; it weighed somewhat over 110 pounds—I forget now whether common or the larger pound unit; for 111 of the larger unit are equal to 259 of the Milanese scale. I saw the stone in the home of Marc Antonio Dugano, which adjoins the sacred temple dedicated to St. Francis. It was of an irregular shape, broken off on every side so as to presuppose a fall. Of an ashy color and rather stained, it gave off when rubbed an odor of sulphur, and was much like a common whetstone. This might have been a misrepresentation, for everywhere such stones are unearthed, and throughout the city are referred to as whetstones.*

I have volunteered this last information, inasmuch as neither in the works of Gaspare Bugatto nor of Francesco Sansovino, careful Italian historians of this whole period, do I find any mention of the shower of stones. Yet why should

*The incredulity of the 18th century scientists regarding meteors was evidently well launched during the Renaissance.

these good citizens have invented such a tale? And indeed, other stones, though smaller, were shown here, there, and everywhere. Nor was this phenomenon to be a favorable omen to the princes who were then reigning, for it is a fact that to men meditating sedition prodigies of this sort add animus for furthering their revolutionary designs, and perhaps are somewhat potent in this respect because they furnish an evident excuse.

Be that, however, one way or the other, during those days it is well known that such a violent earthquake occurred in Venice that the chimes of the churches rang out spontaneously. This ominous event happened in 1511.

In 1513 when Maximilian Sforza, Duke of Milan, the cause of the Principate utterly lost, was besieged at Novara by the French, the dogs of the French came into the city in a pack, licking and fawning upon the dogs of the Swiss. Jacob Motino of Altdorf, a tribune of the Swiss soldiery and a man who had taken part in many engagements, having observed the actions of the dogs, hastened to Sforza assuring him of a decisive victory. This was won the following day.

Here I seem to have exceeded the limits of my theme, but it is pertinent to the present to show how I was born in an epoch in which I was privileged to see many marvels. When I was a young man, and even now, when I am abruptly aroused from sleep, everything in my bedchamber appears to be bathed in luminosity, but soon this force spent itself and faded. It is said that Tiberius had this same experience.

When on the night preceding the 23d of January 1565—the day my man Cesio left and Crasso entered my service—my bed twice burst into flames, I predicted that I should not remain at Bologna. The first time I had occasion to depart, I resisted the desire; the second time, I found it impossible.

In 1552 a gentle little house-dog which had been left at home jumped upon my writing board and tore to shreds my public lectures; the *Book of Fate*, which seemed more in the way of the creature, she left untouched. At the end of the year, quite unexpectedly I gave up public teaching and did not return to it for a period of eight years.

It is ever legitimate to draw inferences from even the most insignificant events, when they are uncommonly persistent, since, as I have elsewhere declared, even as a net consists of meshes, all things in the life of man consist in trifles repeated and massed together now in one figure now in another like cloud formations. Not only through the very smallest circumstances are our affairs increased, but these small circumstances ought gradually to be analyzed into their infinitely minute components. And that man alone will be a figure in the arts, in display of judgment, or in civil life, and will rise to the top, who understands the significance of all these influences, and knows how to heed them in his business. Wherefore in any events whatsoever things that are of apparently no significance ought to be duly observed.

On the day when Lodovico Ferrari came from Bologna with his cousin Luca, the blackbird in the courtyard kept up such an endless and altogether unwonted chattering that we were looking for someone to arrive. Now this was on the last day of November in 1536. Can it be that the latter had anything to do with the former? Not at all! How many times have such omens turned out vainly! By some—Augustus, for instance—forebodings have been observed to their advantage through some incalculable reasoning; by other—as Cæsar and Sulla—they have been held in contempt. It is all like trying to calculate one's chances in gambling: the system comes to naught or is ambiguous. Whatsoever things are above the natural do not subject themselves to natural laws, while such things as may be subjected to natural law contain nothing to excite the wonder of any save the ignorant.

The premonition I had on the occasion when I almost drowned in Lake Garda was quite another matter; I had feared to embark, yet knew not why, so serene was the air.

In this same year many extraordinary events occurred, some of which foreshadowed liberation, others catastrophe—as the breaking of the chain, for instance, by which an emerald had been suspended from my neck. But before that, three rings which I was wearing on my finger—a most amazing thing to be sure—fused into one. Thus the disunion and the union were not a little worthy of admiration, more especially since both a liberation and a condemnation followed close upon these auguries.

However, these dispensations are from God, and, as far as prodigies are concerned, are therefore no marvel.

From my earliest childhood I seemed destined to an early death because of a laborious breathing, an intense coldness of my feet until midnight at least, a palpitation of the heart at that tender age, and profuse perspiration; the latter an uniformly copious discharge of urine carried off in later years. My teeth were few and weak to an exceptional degree; my right hand was not well knit; the life-line was very brief, irregular, interrupted and branching, and all the other principal lines were almost hair-fine or sadly awry. There were stars which threatened, from every aspect, my death, which all declared would be before my forty-fifth year—all vain findings, for I live, and I am in my seventy-fifth year! It is not the fallibility of the art; it is the inexperience of the artificer.

Verily, if indeed there be any such thing as an omen, it could not be more evident anywhere than in Aristotle; and in his works nothing of the kind is to be read.

But let us proceed to the story of my son which in some respects is more worthy of consideration. My son was slain. Within one hundred and twenty-one days Senator Falcuzio himself died exclaiming that he was perishing as a result of the cruel ignorance of a certain one who had persuaded him to vote for the death sentence of my son and had been opposed to leniency. Yet Senator Falcuzio was, in other respects, a man of high character. Him, Hala, who had sickened immediately after the condemnation of my son, followed to death. He was taken with consumption, and expired after having coughed up one of his lungs. The judge, Rigone, who had presided at the trial laid away his lawful wife without a candle to honor her burial—a thing to wonder at, yet true, for I heard it from many. It is also reported that the judge himself, though a man of good name, escaped, by death alone, an indictment which had been brought against him. Even his only son, a growing boy, was swept away to death, so that one well might say that, so accursed, the whole house had fallen under a spell. Within a few days my son's father-in-law, who had procured his death, was cast into prison; and finally, after he had lost his office of collector of debts, he had to beg. A son of his, one he especially favored, ended his life on the gallows, condemned, as I learned, in Sicily.

Of all of those who brought accusation against my son not one escaped without some terrible calamity—being either smitten or destroyed. Nay even the ruler and prince, in most respects a magnanimous and gracious man, after he had become indifferent to my son's cause, on account of some grudge he bore me, or because he was overwhelmed by the number of accusers, was assailed with all manner of evils: he suffered from grave diseases; his granddaughter was murdered by her own husband; he became involved in burdensome litigations. On the heels of these trod public disgrace when the Island of Djerba was lost and the royal fleet put to rout.

Hardly am I so immodest or mad as to think any of these events relates to my own concerns; yet men of the best intentions go down like the grain-fields in a storm, and fall in the times of great calamity when they are deprived of the protection of the most illustrious princes because the latter are involved in public or private misfortunes. For such occasions impious intriguers regularly wait, since nothing offers higher hope of gaining their nefarious ends.

42.

POWERS OF FOREKNOWLEDGE IN MY ART, AND IN OTHER MATTERS

WHATEVER cause—to pursue my story—brought me more of a reputation than I desired in this matter of foreknowledge, whether a divine afflatus, or my harpocratic turn of mind, or a certain perfection of judgment and intellect, I cannot precisely say. In the art of medicine my powers first became conspicuous in the treatment of Cecilia Maggi, in the cases of the son of Giangiacomo Resti, and of so many others, that for so long a period as I was successful no one was able to boast that he had had insight. Rather, even those who were endeavoring to disparage in other respects my methods of medical practice, in this question of discernment always left me first place, though I never claimed it for myself.

Furthermore, passing over other considerations, did I not, while in Bologna, make an offer to the effect that if anyone should be willing to pay ten crowns on the part of a sick person, I would examine the patient carefully twice or thrice, or even but once, and, failing to make an accurate statement regarding the cause of his impending death, I would repay ten times my fee upon the confirmation of my error? Autopsies were held over a number of prominent men, first, to be sure, in my presence, but at length when it was seen that I was never in error, my opponents made secret examinations lest it should be their fate to blush too often because of their ignorance. Not once in the eight years I was professor of medicine there did an occasion for diagnosis arise wherein the others dared contradict me or even voice their complaints, so fortunate have I been in this respect.

Outside my professional skill in diagnosis, anyone might be amazed, surely, at the predictions I made in the consultation regarding Edward VI, King of England,[1] how I discerned what calamities, and from what quarter, were threatening. The foreknowledge of certain events which it is my wish to include in the Song of Lamentation for the death of my son, I pass over in silence, since the events themselves make reference to them impossible[2]; yet I deem it nearer to miracle than to oracle that I foresaw coming events as far in the future as the eighth year after his death.

As I have reiterated, nevertheless, I do not count this prophetic power a tribute to my own glory, and I had rather it came to naught than that I should seek personal honors therefrom. From the beginning I declared Cyprus would be lost,[3] and I gave my reasons therefor. Even about the Citadel of Africa I was by no means uncertain.[4] Yet I would not have anyone think that these things are far-fetched or of a demon, or of astrology, but they are based on Aristotle's idea of prophecy.[5] True divination, says he, is an endowment of the prudent and the wise only.

I was not in the habit of making any pronouncement or prediction before I had inquired into all things which bore upon the situation: I had first learned the nature of the places concerned, the customs of men, the outstanding character of their princes, their history and their noblemen, and had unraveled the negotiations of a great number of their conclaves. From these data, and aided by a system of my own—which I shall not set forth—I was wont to make public my opinion.

Hear, furthermore, what the principles of my system were and what their nature: *doctrina crassa*, *dilemma*, *Τρόπος*, *amplificatio*, my exceptional intuition, and *dialectics*[6]; to these principles I have given long, painstaking and assiduous exercise, and meditation even greater than the exercise.

Certain things have happened to me, for all that, of which

I can render no adequate explanation. By no one had I, in my young manhood, been recommended as a chiromancer to a certain Giovanni Stefano Biffi, yet he, nevertheless, requested me to make some predictions regarding his life. I told him that his companions would play him false, and sought that he might pardon me if I predicted anything serious; whereupon I prophesied that he was threatened with the danger of being hanged shortly. Within a week he was arrested and subjected to tortures; he obstinately denied the charge, but in spite of his assertions of innocence ended his life on the gallows within six months, but not before his hand had been cut off by his inquisitors.

What came to pass with Giampaolo Eufomia, a youth, and at one time my pupil, cannot claim so much from chance. All happened within a month; the chirograph is still at hand. He appeared to be perfectly sound when on a certain evening I directed that my tablet be brought out. Upon it I wrote that if he did not have a care he would shortly meet his death. Herein I had recourse to neither stars nor strategy. I rendered my reasons and submitted them. Within six or eight days he became sick and shortly after died. Such things seem little short of miraculous to those without understanding, whereas, if a wise man reads what I have written, he will say that I saw what was inevitable, rather than that I miraculously revealed a hidden future.

What of that affair at Rome? There are as many witnesses as there were guests at the banquet. I said, "Did I not think some of you would take it uneasily, I should make a statement." One of the guests said, "Perhaps you would like to say that one of us is going to die." "Yes," I replied, "so it seems, and this year." And on the first day of December one who was called Vergilio died.

Our human concerns, furthermore, stand upon the most insignificant circumstances, and by these circumstances important changes are wrought. I daresay such trifles, and even lesser, determine the course of our lives. I do not speak whereof I do not know. A Frenchman came to me while I lived in Rome at the Ranuzzi House and wished to speak to me in private. I said it ought to be enough if we were out of hearing of the others, and when I persisted he went away. Suspicious, I sent some men to seek him out, but no one was found who had seen him. What would you think of such an incident? That man was meditating an evil deed!

Need I explain my predictions about Cyprus? More than once as I heard what were the respective equipment of the Turks and of the Christians, I declared we should have to fear lest we be vanquished. Cardinal Sforza[7] is a witness of this statement. I arrayed my reasons, and the outcome proved what I had said—that the island would be lost through violence and miscalculation.

These things and things like them occur to the studious, the painstaking, and the knowing, not, however, without occasion and not under all circumstances; for those predictions are most reasonable which concern the development of the arts, such as the craftsman's art.

43.

THINGS ABSOLUTELY SUPERNATURAL

THIS INCIDENT happened while I was studying in Pavia: On a certain morning before I was actually awake, I was aware of a knock upon the wall; the dwelling which adjoined the place was vacant; as I was getting awake, and again, immediately after, I perceived another blow as of a hammer. Because at evening I learned that my excellent friend, Galeazzo del Rosso, had died at the same hour—about whom I have had so much to say—I do not ascribe this incident to a miracle. First, the whole matter, owing to the way one sound followed the other, may be ascribed to a dream. Again, it can easily have proceeded from some natural cause, as from a puff of steam. Third, when those I met saw that I was struck by the extraordinary incident and, anxious lest it portend evil, had remained at home the entire day, they made up the story of Galeazzo's death at that hour, and although he had passed away much before, placed it at daybreak of that day, an hour when few who are victims of disease are liable to die. Wherefore I do not place this token, which in so many ways fails to be convincing, among the miracles; but since similar incidents have happened, each person must decide concerning such an occurrence, what seems better, from his own point of view.

In the year 1536, while I lived by the Porta Tosa, it happened that in the month of July—unless I am mistaken—as I was passing from the dining-room into the court-yard, I perceived a strong smell of candles but lately extinguished. Somewhat alarmed I summoned the boy, asking whether he noticed anything. He, since he understood that I was referring to a sound, said he did not; but when I made clear that I was not concerned about a sound but wished to know whether he smelled anything, he exclaimed, "What a fearful odor of wax!" I said, "Be silent," and when I questioned the maid and my wife, all showed themselves surprised except my mother, who sensed nothing, prevented, I believe, by a cold in the head. Asserting that by reason of this portent, a death was imminent, although I retired to my couch, I was unable to sleep.

And, lo, another portent, still greater than the first. On the public street the sound of grunting swine, although none was there; and, thereafter, a similar case, quacking of ducks. What does this mean, I asked myself? Whence these many omens? And the ducks—why have they fore-gathered with these swine who have kept up their grunting the livelong night?

In the morning, bewildered by these many portents, I knew not what to do. I took a walk beyond the city after breakfast, and as I was returning home I saw my mother who beckoned me to make haste, saying that my neighbor Giovanni, at one time overseer of the pest-house, had been struck by lightning. The rumor was that when he vacated this post twelve years earlier, because the plague was raging so violently, he had stolen many things. He kept a mistress, and he never went to confession; and no doubt he had committed other still more grievous sins. Still, he was my neighbor, as there was only a little shack of a house between his and mine; and when I saw and knew that he was indeed dead, I was, by his death, freed from the attentions which, as a neighbor, I had shown. You will ask how the event concerned me. I was spared; I might have been there myself, for, sometimes, though infrequently, I had happened to sit under his portico with him for the pleasure of chatting; it was a cool little nook.

A second omen occurred while my mother lay in her last illness. I had been awakened, and though the sun, now well up, shone brightly and I could distinguish clearly, yet I saw nothing even while I heard fifteen beats—for I counted them—as of water falling drop by drop on the pavement. During the preceding night, moreover, I had counted about one hundred and twenty; but I had misgivings, since I perceived the sound came from the right, that one of the domestics might be mocking me in my anxiety; or that the sounds by day would have passed unnoticed, whereas, occurring by night, they worked upon my credulity. Shortly after, I heard a thumping as of a wagon-load of boards being unloaded all at once above the ceiling, while the bedchamber trembled. My mother, as I said, died; what the noises signified, I know not.

I shall not go into an incident which occurred about the middle of June 1570. I seemed to be up walking about in the night; the doors were closed and the windows barred; it seemed as if I sat down, whereupon there was a great cracking as if issuing from my strong-box. Perhaps this can be attributed to too much imagination. There was no one else whom I could question about the matter.

Who was he that sold me that copy of Apuleius in Latin when I was in my twentieth year—if I'm not mistaken—and forthwith made off? I, who, for a truth, had been only once, up to this time, in an elementary school, and had no understanding of the Latin language, yet had had the folly to buy the book because of its gilded decorations, on the following day found myself as proficient in Latin as I am this very day. Almost at the same time I acquired Greek, Spanish, and French, but merely a reading knowledge; I was altogether ignorant of how to converse or carry on extended discourse in these languages, or of the rules of grammar.

In the year 1560 in the month of May, when, on account of anguish for the death of my son, I had lost sleep little by little, and when neither fasting, nor the lashings with which I tortured my limbs by riding through the underbrush, nor chess, with which I whiled away the time in company with that agreeable youth Ercole Visconte—and even he was becoming exhausted with late hours—distracted me from my grief, I besought the Lord to have pity upon me. For of these endless vigils it was inevitable that I should die or go mad, or at very least give over my professorship. And if I should resign, I had no means whereby I could earn an honorable living; if I became insane I would be an object of mockery to all, I should dissipate what remained of my patrimony, and not a gleam of hope would show a way by which I could hope to change my condition—for I was an old man. And so I prayed that I might die, for this is the ultimate fate of all; and with that I retired to my bed. The hour was late, and I was obliged to rise at four o'clock. I had been permitted to rest quietly in my bed scarcely two hours, when sleep swept me away, and I seemed suddenly to hear a voice drawing near through the shadows; yet whose it was, or who might be there, I could not distinguish because of the darkness. It was saying: "*Why do you complain?*" or "*Why do you mourn?*" Then, without waiting for a response, it went on: "*Because of the death of your son?*" To this I replied: "*Can you doubt it?*" Thereupon the voice said: "*Place the gem which you have on a chain about your neck in your mouth, and as long as you keep it there, you will not remember your son.*"

Awakened by this dream, I kept thinking: "*What has this emerald to do with forgetfulness?*" But after a time when there was no other hope of escaping my misery I called to mind that verse: *Who in hope believed against hope . . . and it was reckoned unto him for righteousness*—[1] speaking of Abraham. I placed the gem in my mouth, and lo, a thing occurred beyond all belief; for at once every memory of my son was lost in oblivion—first, at that time, when I was borne away as into a dream, and then during the future for a period of almost a year and a half, until I wrote the book *Theognoston* or the second *Hyperboreorum.*[2] In the meantime, when I was eating or lecturing, since I could not conveniently benefit by the virtue of the emerald, I was tormented to a very sweat of death. Thus, gradually, I recovered once more the power to sleep and my wonted course of life, as it seemed to me. A feature of this extraordinary incident was, furthermore, that in changing from one condition to the other—from memory of my griefs to forgetfulness thereof—there never seemed to be an instant's delay interposed.

On the night before the 13th of August 1572, there was a light in my room, and I was wide awake. The second hour of the night had barely passed, when lo, I became aware of a terrific noise from the right, as if someone were unloading a wagonful of boards. I looked back quickly, for this was in the entrance of my bed-chamber, or coming from the bedroom where my servant was sleeping; through the wide open doorway I saw a farmer entering the room. I kept looking at him intensely, for many reasons; and he, thereupon, when he was almost on the very threshold of the door uttered these words: *Te sin casa*; and having spoken, he vanished. I neither recognized his voice nor his features, nor was I able to interpret the meaning of what he said by means of any language whatsoever. Elsewhere I have replied to the question, as to why such things happen. Some one may bring up this objection, why marvels of this sort happen to so few? And why, I reply, if thus it is, do men strive so zealously for governments and magistracies, or to realize their ambitions by so many abominable means? I aver that this is not the place for determining these matters, nor are my shoulders capable of bearing such a burden, but it should be turned over to the theologians. Let it be enough for me to have told the truth of my tale.

I shall forbear to relate the incident of the thunder at Bologna which, because it reverberated above my bed-chamber without a tremor, was less foreboding—like that ever accursed clatter of boards. Yet after these alarms, no death ever followed, except my mother's, but she was dying of disease, weakened by old age.

Likewise, I may pass over the details of the obstinate conduct of my clock. This can easily be explained by natural causes, as can the incident of the earth which, day after day, in October and November of 1559 was thrown out from beneath my hearth, almost in plain sight. This I saw with my own eyes, nor was I sleeping, for it happened in clear daylight.

About the 25th of March 1570, I had written down a prescription for my patron, Cardinal Morone, and was annoyed to find that one page of this had fallen to the ground. I arose, accordingly, and even as I stood, that page lifted itself in the same moment, and passed across to the wainscoting where it clung to the panels in an upright position. Moved to amazement, I summoned Rodolfo and showed him the marvel; but he, however, had not seen the peculiar movement of the paper, nor was I able to fathom the meaning of the portent, since I would not look for evil at every turn. It turned out that all my affairs were changed, and somewhat milder breezes blew.

A month later—I believe in June—while I was writing to the same man, I was seeking my letter-sander. When I had hunted in every possible place for some time, I picked up the sheet of paper that I might gather a little earth with which to sand it; and thereupon I saw the sand box had been hiding under the paper—a little box only an inch and a quarter high, and round, so that the diameter would be but an inch. Yet how did it happen to be hiding from me under the very paper upon which I had been writing, as on a level? The incident had this effect upon me: it strengthened the confidence I had conceived in the philanthropy and wisdom of my correspondent, which wisdom I trusted he would employ in my behalf with the most excellent Pontiff[3] to the end that I should no longer be obliged to suffer such adversity in view of my long labors.

But what followed on the 9th of October of the same year made all things seem fair, plain, and clear. I had been imprisoned on the 6th, and had given bond, 1,800 gold crowns. On the 9th, at the ninth hour of the day, the sun was shining brightly in the apartment in which I was imprisoned. When those who had arrested me had gone away, I bade Rodolfo Selvatico to fasten the door of the bedroom. This he was loath to do, and was exceedingly surprised at my absurd idea—whether because God so willed, or whether I was influenced by reason—that I, because I had been forced to endure the ignominy of incarceration, should show no willingness to mingle voluntarily with my jailers. However, he obeyed, and lo, the door of the chamber was no sooner closed than a terrific blow struck it, resounding near and far; the next instant, even as we were gazing, the shock leaped across to the window frame, on which the sun shone in a dazzling light, with a similar percussion, and there rattled against the window and the lattice, as if the fastenings were creaking. Then it ceased.

When I had observed this portent, I began at once to lament my miserable fate, but it became evident that what I was interpreting as a sure sign of a wretched death, pointed toward life. Not long after I began to reason with myself in this manner: If so many princes, although young, strong, and happy, expose themselves to certain death for their kings, that they may win their favor, when they can hope for absolutely nothing from their dying, how is it possible for you, an old man, wasted, and almost dishonored, to suffer, either for your sin, if they will judge you guilty of it, or for the injustice done you, if you be unworthy of this evil in the eyes of God, who by his goodness shows that all your affairs are under his care?

And thereupon I seemed safe from death which I had been fearing, and I led a pleasant existence in so far as our human nature permits. In this manner, then, I lived—though all was over, as it were, with my life—while I remembered that previously I had not been able to endure even a moment in confinement without seeming to suffocate.

Rodolfo was, as I have said, present when this curious incident occurred. He took his degree the next year.

Yet such unnatural occurrences have this effect, that while they are in the very act of happening or shortly before they come to pass, they make a great impression on a man, engaging his whole attention; when, however, the first heat of their influence has grown cool and the effect is diluted, unless you have made sure of their reality by nailing them, as it were, upon your consciousness, you almost doubt whether you have actually seen or heard anything. And this, I believe, happens for reasons even deeper than the gulf between our own natures and the causes which produce the phenomena.

I know certain men who, in order to appear satirical, scoff and incite derision in regard to any such apparently supernatural occurrences. The prince of these is Polybius, the philosopher, without a philosophy, who did not know what the duty of an historian was, but, by going beyond his province, made himself ridiculous, although at times he is worthy of admiration as when he discussed the Achæans in Book II of his *Histories*.

What more? Tartaglia was right when he said nobody knew everything; nay, that all those have not the beginnings of understanding, who do not realize that they are ignorant on many points. Look at Pliny: although he published a very excellent history, he shows himself an ox when he treats of the sun and the stars. Is it surprising, then, that Polybius, inasmuch as he is concerned with loftier and diviner themes, so patently betrayed his crudeness?

One thing alone is sufficient for me: to understand and grasp the meaning of all these wonders would be more precious to me than the everlasting dominion of all the universe; and this I swear by all that is holy.

If I had wished, I might have included in this chapter the wonders which my parents used to relate to me in the likeness of fables, and which then seemed worthy of a smile. Later, however, did I not realize all too surely that these very things might seem insignificant? For I have not dared to attribute so much to the wisdom and diligence of my elders that I could hope their accounts of marvels could be confirmed.

In truth, I have this to give me assurance enough, that such wonders are unexpectedly seen in the presence of those about to die who have been signally good or evil. Wherefore, this being the case, these omens cannot be so much fortuitous, as natural or divine. For in the mind, which is neither paralyzed by fear nor unsettled by conflicting emotions, violent spiritual stimuli, accompanied by demonstrations of a miraculous nature, undermine the foundations of false notions rather than increase them. . . . How would a miracle affect a girl praying to God at the bidding of her father, whom she thought ought to be freed?[4]

But enough of these matters! For I have indicated herein, as briefly as possible, simply when such incidents occurred, what was their nature—incidents from which, moreover, all suspicion of either error or fraud was absent. I have passed over a great host of others which were not sufficiently pertinent, although they were plainly manifest; I have omitted events of whose extraordinary character I am even more convinced, since they occurred in my presence, yet were not so rich in testimonials of supernatural power. Everyone may learn these from my *Commentaries*.

This alone I ask you, O reader, that when you peruse the account of these marvels that you do not set up for yourself as a standard human intellectual pride, but rather the great size and vastness of earth and sky; and, comparing with that Infinity these slender shadows in which miserably and anxiously we are enveloped, you will easily know that I have related nothing which is beyond belief.

44.

THINGS OF WORTH WHICH I HAVE ACHIEVED IN VARIOUS STUDIES

LET ME REMIND you, by way of preface, that there is practically no new idea which one may bring forward: In *dialectics*, although one system had been known—the Aristotelian —I have divided the system and the practice of it so that individual disciples may make a study of the precepts of dialectics according to Euclid, Ptolemy, Archimedes, Hippocrates, Galen, and Scotus. In addition I have enlarged upon the practical application of *dilemma*, and, in like manner, of *doctrina crassa*, of the trope, of elaboration, and analysis; by means of these doctrines very many men strive eagerly to discern the incorporeal form, and separate, as it were, the soul of things from the physical structure; putting thereby experiments in casuistry, which excite amazement, before true scientific knowledge, so that out of a limited field of experience they come to far-reaching conclusions. It is as if among mortals they would complete the cycle, and bring together beginnings to ends, even as it is with the immortals. Neither is it a part of their system to select figures or examples from readings—much less accurately copied references—so that they reduce the scholarly labors of many months to an hour's argument.

Even further does their superficiality go: the gift of extemporaneous lecturing which at another time they made much of, now they cease to respect properly. And those men are to be pardoned who lay such powers to the influence of an evil genius, since they recognize neither what is really a good, nor yet the favor of God.

In arithmetic I advanced almost the whole field of the

science including the sections treating, as they call it, of algebra[1]; my discoveries dealt also with properties of numbers, especially of those having similar ratios between themselves. I also expounded the numerical functions already discovered, showing either a simplified treatment or some uncommon formula method, or both. In geometry I dealt with confused and reflex proportions, and the treatment of infinity with finite numbers and through finite, although it was first discovered by Archimedes. In music I discovered new tones and new intervals, or rather brought back into practice and use such tones and intervals as were already found according to the treatises of Ptolemy and Aristoxenus.

In natural philosophy I withdrew fire from the number of the elements and showed that all things were essentially cold; that the elements were not reciprocally changed; and I upheld the doctrine of palingenesis. I demonstrated that there were only two true qualities—heat and moisture. I set forth the essential qualities of salt and oil. I proved that the principle which underlies the generation of perfect creatures consists not in the act of coition unless that be attended by heat of atmospheric origin.

I taught that God ought to be called infinite; that all things which possess differentiated and organized parts have a life-principle; that the existence of our own life-principle and its immortality is, according to the philosophers, real, not a shadowy dream; that all things go by numbers—as, for instance, in one variety of plant there are the same number of leaves in a group and the same number of seeds.

I demonstrated that the principle of analogy consists in a process active through one medium and one material, and that from this results so many varieties and so much beauty. The earth is a thing of itself and not, as it were, mingled with water, wherefore it happens that often the one thrusts itself out into the opposite lying parts of the other. Why is the East better than the West? Why, when the sun has turned back after the solstices, the extremes of temperature may be increased for many days? What is Fate and how does it control human affairs?

I have dealt with the causes of certain singular circumstances, as why a thousand dice in a thousand throws, if the dice are not loaded, necessarily cast one ace.[2]

I have shown that from all leaf mould of plants, creatures of quite different genera are produced according to the nature of the leaf; that nature is of itself nothing, rather an imaginary and void conception, introduced, to become a beginning of many errors, by Aristotle, to the sole end that he might have the name of overthrowing the philosophy of Plato.

Other doctrines too numerous to mention I have propounded, but this especially:

I have taught that a contemplation of the natural universe leads to an artistic expression and to creative labor, although no others before my time have undertaken to offer such a suggestion.

In the province of moral philosophy I have declared the equality of the condition of all beings, not only of all men but of all living creatures; whence it is permissible to infer an equalizing of all activity by death; and among men the propriety of seeking advantage from adversity. I have inquired also as to what way of life is best of all, and how that life can maintain an even tenor. There are three states in the realm of moral philosophy; to be ignorant of what good and evil is, when one is not aware of the significance of either, is frequently desirable in the common concerns of human existence. Again quite the contrary is true; under still other conditions it is a choice or question as to what constitutes good or favorable action in any given circumstances. An understanding of the *mores* of men in general is an important consideration; after that, an understanding of them according to racial traits, and any other characteristic, such as the customs, finally, peculiar to them generally or racially.

To the art of medicine I have contributed the true reckoning of the duration of the crisis in diseases; a general relief for gout and for pestilential fevers; the manifold transmutations in oils; how from medicines not purgative to prepare cathartics; a treatise on the special properties of waters.[3] I have discovered many a useful method of preparing food; also a way to convert medicines of little medicinal value into useful and ready remedies, and such as are extremely distasteful into remedies easily administered. I have a remedy which so efficaciously relieves and strengthens a person suffering with dropsy that he may go about his business in the city on the same day.

This likewise is my discovery, how from a comparison of the cures of one member and another, some understanding of the causes of the disease and method of treatment may be deduced; and I have shown how in a repeated study of one book—three or four readings—an understanding and method of treatment of divers diseases may be obtained. The true and shorter procedure of operating for hernia was called back into common usage through my efforts. I wrote a trustworthy history of urinary diseases, since we possess only certain obscure remains of these studies.[4]

A commentary on the most difficult books of Hippocrates has been prepared by my hand, particularly of those books judged authentic. The work is not yet completed on this day in which I write these words, that is, November 16, 1575.

Besides these, I have collected a mass of material on the *mal français*, and brief results of experiments pertaining to several diseases very difficult to treat, epilepsy, insanity, blindness. My findings, for instance, concerning the use of equisetum for dropsy is included in this collection. I have elsewhere set down my numerous discoveries pertaining to scirrhosis, to excessive desire to urinate, to diseases of the joints, to gravel in the kidneys, to colic, hæmorrhoids, and other diseases—about five thousand suggestions for treatment, all told. Of problems solved or investigated I shall leave something like forty thousand, and of minutiæ two hundred thousand, and for this that great light[5] of our country used to call me "The Man of Discoveries."

45.

BOOKS WRITTEN BY ME; WHEN, WHY, AND WHAT BECAME OF THEM

PUBLISHED WORKS

Treatises on Mathematics
The Great Art · 1 Book
On Proportions · 1 Book
Regula Aliza · 1 Book

On Astronomy
Commentary on Ptolemy · 4 Books
Selected Genitures · 1 Book
On Interrogations · 1 Book
On the Seven Planets · 1 Book
On the Use of the Ephemerides · 1 Book
On the Emendation of Stellar Movements
and Recognition of the Planets · 1 Book
Encomium of Astrology · 1 Book

On Physics
On Subtlety, with a Defense[1] · 22 Books
On a Variety of Matters · 17 Books
On the Immortality of the Soul · 1 Book

Moral Discourses
On the Uses of Adversity · 4 Books
On Consolation · 3 Books
An Exhortation to the Noble Arts · 1 Book

First Short Works on Various Subjects
On My Own Books · 1 Book

On Marvelous Cures · 1 Book
In Praise of Nero · 1 Book
In Praise of Geometry · 1 Book
First Treatise on Hidden Knowledge · 1 Book
On the Essence of One · 1 Book
On Gems and Colors · 1 Book
On Death · 1 Book
Tetim or On Mortal Condition · 1 Book
On the Good in Small Things Near at Hand · 1 Book
On the Summum Bonum · 1 Book

Second Short Works on Various Subjects
Dialectics · 1 Book
Hyperschen[2] · 1 Book
On the Zeal of Socrates[3] · 1 Book
On Water · 1 Book
On the Ether · 1 Book
On Medicinal Potions · 1 Book

Third Volume of Short Treatises on Medicine
On Causes, Indications, and Seat of Diseases · 1 Book
Small Treatise on the Art of Healing · 1 Book
First Book of Medical Advice · 1 Book
On the Malpractices of Doctors · 1 Book
Proof that No Uncompounded
Medicament is Harmless · 1 Book
Triceps[4] · 1 Book
A Defense of the Physician of Thessaly · 1 Book
A Defense of Camuzio · 1 Book

Commentaries on Medical Works[5]
On the Aphorisms · 7 Books
On the Composition of the Air · 1 Book
On Poisons · 3 Books
Total · 11 Books

On Prognostics · 4 Books
On Seven-month Parturition · 1 Book
Total · 5 Books

On Air, Waters, and Localities · 8 Books
Second Book, Medical Advice · 1 Book
Total · 9 Books

On Nourishment · 1 Book
Examination of Twenty-two Cases of Sickness[6] · 1 Book
Total · 2 Books

On Methods of Divination
On Dreams · 4 Books

Published Works Not Included in the Above
On Wisdom · 5 Books
Antigorgias[7] · 5 Books
In Praise of Medicine · 1 Book
Supplement to the Almanac · 10 Books

MANUSCRIPTS[8]

Mathematics
New Geometry · 2 Books
On Whole Numbers · 1 Book
On Fractions · 1 Book
On the Properties of Numbers · 1 Book
On Surds · 1 Book
On Imaginary Expressions · 1 Book
On Music · 1 Book

Physics
On Nature · 1 Book

Fourth Book on Hidden Knowledge · 1 Book
Of the Hyperboreans · 2 Books

On Morals
On Morals · 3 Books
On the Best Way of Life · 1 Book
A Memoir[9] · 1 Book
The Book of My Life · 1 Book

Medicine
On the Urine · 4 Books
On Living Conditions in Rome · 1 Book
On the Teeth · 5 Books
On Safeguarding the Health · 4 Books
On the Indic Plague · 1 Book
Third Book of Medical Advice · 1 Book
Actus · 1 Book
On the Contradictions of Doctors · 12 Books
Handbook · 4 Books
On the Treatise: Regimen in Acute Diseases · 6 Books
On the Ars Medica of Galen[10] · 1 Book
Floridorum on Hasen's Doctrines[11] · 2 Books
On Epidemics of Hippocrates · 5 Books

Theological Works
Hymn and Life of the Blessed Virgin
Life of Saint Martin with notes

On Various Arguments
Paralipomena · 6 Books
On the Books of Famous Men · 1 Book
On Discoveries · 1 Book
Problemata · 1 Book
On Writing Books · 1 Book

Proxeneta · 1 Book
On Games of Chance · 2 Books
Dialog on Imprisonment · 1 Book
Flosculus, a Dialog · 1 Book
On the Nodes · 1 Book
Antigorgias: a Dialog[7] · 1 Book
Encomium of Medicine · 1 Book
Metoposcopy · 7 Books
On Clever Devices · 1 Book
On the Use of the Ephemeris or
of New Discoveries · 1 Book
Sacred Things · 1 Book

The reason I was induced to take up writing I think you have already learned. I was, in fact, urged by a dream, and thereafter twice, thrice, and four and indeed many times the suggestion was thus presented, as I have elsewhere made known. But I was also urged by a great longing to have my name live.

On two occasions, furthermore, I have lost a great mass of writings and a number of books. First, around my thirty-seventh year I burned about nine books because I knew they were empty of content, and would never be of any use. I had accumulated, in the meanwhile, a hopeless farrago consisting of medical writings chiefly. I did not remove any parts from these books nor preserve any entire except the manuscript of the *Book on Poor Methods of Healing in Present Use*,[12] from which I had begun the publication of a first edition, and the manuscript of the rudiments of arithmetic, from which I composed a small work on arithmetic.[13] Shortly after, about 1541, I enlarged and published anew a little tract, a *Supplement to the Almanacs*, which I had already had printed once. In 1573, when that disastrous occasion[14] had at length passed over, I burned another 120 books, but not as on the first occasion. For this time I selected and removed from the books whatever seemed worth anything to me, and in addition, I saved some entire, such as *The Book of Clever Devices from Narrative Writings* and the *Book on the Writings of Famous Men.* In other works I made changes like Diomedes who exchanged *Χρύσεα χαλκείων ἑκατόμβοί ἐννεαβοίον....*[15]

This I did because the latter continued in my favor; I destroyed the former because they displeased me, and the outcome was favorable to both courses.

After I had been prompted by a recurring dream, I wrote my books *On Subtlety*, which I enlarged after the first printing, and then, having included even more new material I had them sent to press for the third edition. From this I turned to the *Book of the Great Art*[16] which I composed while Giovanni da Colla was contending with me, and also Tartaglia, from whom I had received the first chapter. He preferred to find in me a rival, and a better man at that, than an associate bound to him by gratitude, and of all men most devotedly his friend.

While I was traveling down the Loire River and had nothing to do I wrote my *Commentaries on Ptolemy* in the year 1552. I added *On Proportions* and the *Regula Aliza* to the *Ars Magna* in 1568 and published the work. Following this I also corrected and had rewritten the *Treatise on Arithmetic*, two volumes entitled *New Geometry*,[17] and a work on *Music*,[18] but this last I corrected and revised six years later, that is in 1574. The several volumes *On a Variety of Matters* I published in 1558; they consisted of the remaining material of the work *On Subtlety*, which I was not able to put in order nor whip into shape on account of the host of cares which beset me. For my sons were little inclined to obedience, or conformity; my financial returns were next to nothing; the demands of reading gave me no respite; what with domestic arrangements, the practice of my profession throughout the city, prescriptions, letters, and so many other distractions, there was not time to breathe, much less any opportunity for revision of my writings.

On Consolation I published for the first time in the midst of the sorrows which beset me. Later I added *On Wisdom*, so that the whole might be printed for the second time in the year 1543. In the meanwhile I wrote those numerous short treatises, some of which are published, others not yet, and I also wrote the whole body of my medical works, four of which you see published—*On the Aphorisms*, *On Nourishment*, *On Air, Water, and Places*, and *On the Books of Prognostics.* Thus far there are ready two commentaries entitled *Floridorum*, a commentary on the *Ars Medica of Galen* and the first and second parts of *On Epidemics of Hippocrates.*

When I came to Bologna *On Dreams* was published, a work destined to be useful no doubt to many men of understanding, but perchance a source of detriment to the uncultivated plebeian; but what is there which, if anyone use it wrongfully or inadvisedly, may not prove harmful? Horses, swords, arms, munitions in the hands of violent men are all terrible weapons; to good men such things are not so much conveniences as necessities. Any sort of division of my theme, as between profitable material, that not especially useful, and a third section to be read by the learned alone, offered difficulties.

I wrote *Dialectics* in order that I might show how to give a material form to those shadowy precepts which underlie the abstractions of this doctrine. Gratified, then, with my own work, by reason of the pleasure it afforded me, I had it published, although it is, however, to be considered neither complete nor even fully emended.

My *Small Treatise on the Art of Healing* I gave out for public use when I noticed that the rest of my writings were coming forward rather slowly. The book *On the Immortality of the Soul*[19] I wrote more for the purpose of a study of the subject than as an expression of final opinion, and since this did not do justice to the great amount of material involved, the second book of *Hyperboreans* took the place of it.

I composed the *Dialogs*, one for assuaging the pains of adversity which had overtaken me, the other for assailing the madness of men, so that by four contrary meditations I might ease the prick of grief, mad pleasure, stupid greed, and fear.

Proxeneta[20] was the result of an impulse to express myself; in the writing of the *Memoir* I distinguished myself.

Four volumes, in the nature of a compendium or thesaurus, include, with brief explanation, the flower and fruit of the whole medical art, so that you never need lack for want of a single point, and if you will have consulted earlier manuals rather than these books, your labor will be as good as lost.

I wrote my commentaries *On the Books on Regimen in Acute Diseases*, so that, based on a trustworthy doctrine, the regimen of those who can be saved in acute attacks of disease may be maintained. In curing diseases of this sort I have been, as I have said, exceptionally lucky.

The *Treatises on Urinary Diseases* are not yet complete. These demonstrate the wonders of nature, since in a subject so lightly esteemed, so many remarkable features are embraced; and so to the universe we must concede the wonders of its parts. The treatment of these books is, nevertheless, simple, and for this very reason they offer the greatest difficulty to imitators. The structure is well worked out, furthermore, and properly verified by many tried propositions.

Since my books, *On the Contradictions of Doctors*, touch upon all the doubtful points of the profession, and, in so far as I was privileged to discredit them, assail doubtful practices, if the sentiment I expressed pleased, why should I condemn the work? If I do not have faith in what I said, why did I set it forth? Why did I pass judgment? Why did I decide to do it in the first place?

The work *Problemata* was written with this in mind, as the saying goes: *Et prodesse volunt et delectare poetæ.*[21]

The book *On Games of Chance* I also wrote; why should not a man who is a gambler, a dicer, and at the same time an author, write a book on gaming? And, peradventure, *ex ungue leonem*, as they say.[22]

Thirteen books of the *Metoposcopy* I condensed into seven; of this the treatise *Physiognomy* is a part. I am indebted to Girolamo Visconte for this. Suetonius exalts this art with extraordinary praises; I have discerned certain shadowy truths in it, but whether they are actually truths, or deceptions is very difficult to determine, for you will be caught by deceptions simply because of the host of men and impressions, and their never-ending mutations.

The *Paralipomena*[23] as they stand at present are the fragments which I preserved from that earlier reorganization of my work in which I sacrificed much for no other reason, it must be supposed, than that I was not wont to pass approval on writing exhibiting confusion and disorder. In assembling the *Paralipomena* no such choice was observed; worthless writing stands with that of highest order; the unseemly with the decent; the useful with the harmful, the carefully revised with the casual, the curious with the absurd—a great farrago. Nor had I hopes, even though I had made away with so many works, of being able to revise what remained and reduce it to one well-rounded whole. I have deemed it better, if, while all see and live and the circumstance is known to all, I should rather look to the interests of my friends and patrons; and what is more, and what entails an enormous saving of time, that it would be far better to be able to leave to posterity more works clearly and accurately written, than to make a show of all I have cast aside.

For a man so to live that his fellows may feel no lack of those duties which he as a man of high character in his profession ought to discharge—this, I aver, is honor's crown. Wherefore I wrote *On Discovery*, *On Writing Books*, and *On the Books of Famous Men* so that, by the deed itself, I might confirm those things my words had praised. The *Hymn* and the *Biographies* I wrote in order to show myself grateful to those from whom I had received so many blessings. The *Annotations* I subjoined because I considered it of the utmost importance that precise care should not only be exercised in the biography, but I felt, indeed, that it was due from me. For as these attentions are the strength and beauty of even the greatest works, similarly errors neglected, and passages inaccurately written, shake the confidence of the readers' minds, detract from the influence of the books themselves, and work harm to the common good. The examples of Aristotle and of Galen have revealed to me the possibility of accomplishing this perfection of execution; to them this had been needful since they treated of themes of universal interest; for me it is a safe and proper policy, inasmuch as I, through careful attention to details, have taken the greatest pains, in proportion to my powers, in the book entitled *On the Best Way of Life*; since, than in such devotion to my work, I found no other avenue of escape from the memory of the past, the evil of the present and the lowering perils of the future. Without such an absorbing passion I question whether anyone in mortal life is able to think of himself as immortal; whether he can die without experiencing the sorrows of old age, and yet outlive his youth; if in the eternal tumult of things he can find peace, and in the endless round of life's vicissitudes, anything constant.

With these four favors I should willingly be content, and should forebear to take into consideration the many calamities and other evils which are far more numerous in the life of every individual than these needful benefits. The greatest of all the calamities, utterly inevitable, which came to pass was the death of those dear to me. But I question whether that had to happen the way it did. Yet what matters it? Does not the same round of vicissitude come to all others? But it might have been deferred. What difference, if it had to come sooner or later!

There never has been, nor will there be, any rest for mortal man. Compare the trials which it has been your lot to endure to the disasters and the state of affairs in the time of Polybius: the present is a wreath of roses compared to the horrors of those times, which can truly be termed calamities. Nothing was safe to men then: murder in cold blood, slavery, and the loss of all men's property to ruthless spoilers was mere pastime.

Another consideration is this: the contemplation of an eternal and blessed life known to us which was not a dispensation to those of other lands and times. What that can be called deplorable is able to overtake one whose thoughts are fixed upon this hope; the same origin, the same common goal, and the same lot, all have shared, except that for us joy is the guerdon for death. Yet we are cheated, for a truth, out of the four favors I have mentioned above: first, because we believe that this existence offers something more tangible than is actually the case, even in the midst of its activities; second, because we fail to observe that there is nothing lasting, not to say eternal, in life; third, because we think the intellect ages even if the vital principle lives on, since we argue that what is common to the body must needs grow old with it. But I forthwith aver that no part of us grows old—not the life principle, if its medium and its motive force have permanence, as I now believe; not the body, because, according to the philosophers, especially the Platonic philosophers—as we read in *Phædo*—it is not the chief part of man; not the active principle, which is the first and fundamental part of man, and is the sum of all perfection, although its operative force can be impeded, not because it lacks in some respect, for its powers are varied. The sun needs the air in order to shine; the air is the cause of the illumination of the sun, yet impurity of the air impedes the radiance of the sun.

The preceding, therefore, was the summary of my counsel and the scope of my work.[24]

There is a second work called *Memoralis,* in which is brought together the sum of the teaching contained in the earlier book. In this second the material is so arranged and grouped according to topics that not only can you find comfort in it, but help for whatsoever condition you have fallen in with.

The third work of this group[25] is a *Handbook* which sets forth not only sentiments on gain and honor, but also considers that it is a most excellent virtue to discharge the duties of piety and to fulfill one's obligations, rather with this attitude, that one should recognize even how much more is required of one at the very point where others think their duty ends. If it is a pleasure to the architect to have built a house not aimlessly but upon a carefully drawn plan, how much more gratifying it is for a man to have been able to render just service to his fellow and recognize how much he can help him.

The fourth book is this very conclusion of all my writings,[26] composed for my own pleasure and out of a sense of duty; and, if I have wittingly violated this by any falsehood, in what attitude would you have me approach my God? Or what pleasure, think you, I should find in my work? Granted, of course, that my case is not the case of a man leading the existence of the brute beasts which take no delight in what is of good repute, but find gratification only in the things they are able to consume—as spiders flies and ostriches bits of iron! For with those who are highly regarded and wish, by a clever deception to appear to know more than others, when, upon being put to test, it is discovered that they are really ignorant of many things known to others, I am unconcerned.

In the fifth place stands my work *On Safeguarding the Health*, and finally, sixth in order, stands the second volume of *Hyperboreans*. In the fifth work the third volume is complete in every detail, and in the sixth, the first volume.

Except for this group of nineteen works I could desire that no other book of mine survive. Peradventure this may seem surprising to some; yet did not Vergil express a desire to have the *Æneid* destroyed—nay, it was his will and command —leaving only the *Bucolics* and *Georgics*? And I, not however before having reviewed all my work, have adopted the same attitude.

I completed the books *On Nature* the reason for which has been sufficiently discussed.

Theognoston I wrote, taking in the *Hyperboreans*; *On Morals* was composed in imitation of Aristotle.[27] In his *Republic*[28] he estimates that the very longest sway of a tyranny scarcely covered one hundred years, although this is an error.

The Book of My Life I was moved to write, because this seemed meet, necessary, and in accord with the circumstances. And it is not unpleasant for me so to review my days, if I may put any faith in the sentiments of Epicurus.

On the Teeth was composed on the basis of methods I proposed as a cure for diseases of long standing by a reliable method, which was exactly the plan I followed in putting together the commentaries *On Cases of Acute Diseases*.[29]

On the Indic Plague was the result of what many had written to me on this subject, and I had condensed the whole confused mass into this treatise.

My reasons for composing *On Safeguarding the Health* were many; in the first place Galen's treatise is too pedantic, and, while he preserves a logical sequence, leaves, I regret to say, many points obscure, and many doubtful and undetermined, inasmuch as in his discussion of massage and exercises he wanders from the point, goes far afield and lingers on irrelevancies. For example, although he had many a good opportunity throughout his numerous works, he neglects to point out what wine might properly be given to those in good health and to young people, but seems rather to have avoided the topic intentionally. I pass over the fact that neither the practice of the ancients nor of the Greeks, indeed, is suitable for Italians and for the present time, nor are the circumstances quite the same. He mentions, for instance, in his second treatise on foods that the Arum is eaten in the country of Cyrene for turnip, without being in any way harmful. He was unacquainted with the process of distillation not yet discovered in that age. The authority, nevertheless, of him whose very high position and virtue have been conspicuous, has necessarily had great influence with me.

The little volume called *Actus* I kept as a handbook in which I briefly jotted down ideas for further elaboration; it is to my thinking as the spark to the cannon.

As for the rest, if anyone should wish to reduce all the remaining books of mine to the norm of the first eighteen, by revising the points which lack coherence or which are indispensable—even as I have done in certain places of other works, for example in the *De Varietate Rerum*—he will find it worth his effort, and I shall be greatly in his debt.

You will remember, moreover, that, since all books, at least all good books have been composed under the radiance of divine inspiration, this nevertheless has been the result of three influences. First, they have been written, as it were, in the light of that force which is common to all, for all wisdom is from God, the Lord, and, as the Platonists think, our intellect is united thereto through the agency of Eternal Good, becomes endowed with the faculty of understanding, and illuminates, besides, the transformed life.

Another method is more manifest inasmuch as the inspiration is imparted by degrees; on this point some Platonists have mistakenly been in doubt; my interpretation of the rule offers no uncertainty, however, and such inspiration is bestowed upon all upright men.

A third source of insight is that which is grasped when certain occasions arise, as that which, in my case, occurred on the 14th of March of this present year 1576 when I was writing, in the book on the care of the health, the history of ferula or fennel-giant, and was praising the use of fennel as an agreeable medicine. An old man, not decently clad but in rags, hurried up to me that day while I was in the vegetable market which is near the fish market. He dissuaded me from the use of fennel, saying: "It can, according to Galen's opinion, destroy instantly, exactly as hemlock." When I replied to this that I made a proper distinction between hemlock and fennel, he said, "Take care, I know what I'm talking about," and he went away murmuring something about Galen.

Having returned home I found a place which, on other occasions, I had not noted. Therefore, although the change did not amount to much, to be sure, I nevertheless altered my former statement, and added many exceptions: namely, the fact that the ferulaceous herbs (considered in my treatise) had been collected in Italy; that whether they were safe for use could be determined by an examination of the stems of these various species; also if they had not been frostbitten; I added that they could, as a natural remedy, though not commonly known, be used, but not abused; that they should be gathered in the early spring and not in the beginning of the hot weather, and in a place where clover is abundant; that if anyone felt any fear of this he should take beforehand some zedoary[30] or medicinal apple, bread baked with garlic, or else try some on the dog or the hen.

Accordingly if anyone give heed to this third source of inspiration for writing he will cull no slight reward from everything he reads and will be grateful, as I have said from the first, to anyone offering a suggestion. And I beg men to offer corrections to my work—*lege Augustana*.[31]

My later works have all been written, as I have just said, with the same technique but not with the same care.

46.
CONCERNING MY OWN EXISTENCE

TO ME MEDITATING upon the foregoing themes and their like, this question occurred, which I feel can be justly put to me: that is, whether, in view of all the evil that has befallen me, all the good that has come my way, and all the ordinary events of everyday life, I am sorry to be alive or regret having lived? It would be foolish not to weigh carefully what one would say, and to one's comments not add the reasons for expressing them. My misfortunes have been the death—bitterest of blows—the folly, or the sterility of my children; my own impotence, my never-ending poverty, conflict, and accusations; troubles, diseases, dangers; my imprisonment, and the injustice of tolerating casualties not deserved, so many in number and so often.

But let us pass over the vicissitudes common to all. If he be not unhappy who has neither children nor honors nor resources, how does it stand with an old man who has some share in all of these? Compare yourself to your origin and to those more wretched than yourself, and not to the more powerful only; when I thus remind myself I cannot rightly bemoan my condition; nay, I am rather fortunate, in view of my exceptional and accurate understanding of many wonderful things, if we are to believe Aristotle. Yet I claim I am happier for this very reason, because I am able like the Stoics, to despise the vicissitudes of mortal existence; and the fruit of this attitude I now realize to be bountiful indeed, so that I am not one who needs envy myself as I was in youth—infirm with age though now I be. Nor is my condition pitiable, inasmuch as I have been allotted the use of my physical senses and a share of the benefits of fortune, and by reason of my understanding. Nay, I shall say even more—that I am one of the most fortunate of men, since I know our own human nature partakes of divinity.

If anyone should announce to a man overshadowed by death that he would survive for fifteen years, as even God to King Hezekiah,[1] will he not go his way more joyfully? Or if from his thirty or even his hundred years he might pass on into a thousand with what boundless delight will he be affected? Will he not consign to oblivion all mortal joys? If this time were to be extended to ten or a hundred thousand years how will he control his ecstasy? But suppose then he be granted an eternity; what more will there be which he may seek or hope for? But he who lives without this hope is deprived of a double and true good—that is, the hope itself and the rewards of it. If therefore it has, after this manner, been God's pleasure to make us in our mortality, participants of immortality, it does not befit us to neglect his free gifts, nor to hold any other than a hopeful view of our condition.

47.
GUARDIAN ANGELS

ATTENDANT or guardian spirits—the Greeks are wont to call these angels, nevertheless in Latin the word is *spiritus*[1] —are recorded as having favored certain men constantly as I have elsewhere noted—Socrates, Plotinus, Synesius, Dio, Flavius Josephus—and I include myself. All, to be sure, lived happily save Socrates and me, yet even I, as I have just said, am favored with the best of circumstances. But Gaius Cæsar the Dictator, Cicero, Antony, Brutus and Cassius, evil though dazzling fates attended. Glorious spirits, yet each fatal, followed Antony and Cicero. Josephus had a high spirit of exceptional nobility, which by virtue of his courage in war found him favor with Vespasian; and under Vespasian's sons his life was eventful by reason of his wealth, deeds memorable in history, three sons, and that struggle with the great calamity of his people. His was a spirit, then, characterized by foresight which took into account the future, and by which he was rendered famous in captivity, liberated from the insane exhortations of his own companions and saved from the depths of the sea.[2] Truly the guardian spirits have clearly been operative, and it has been my lot to be attended by a good and compassionate angel, I believe.

This, I am persuaded, had long attended me, but how it apprized me of perils impending I did not realize until I had completed my seventy-fourth year of life and was in the act of writing this autobiography. Then only was I able to grasp all. That the many threatening events, hovering as it were, on the very threshold were accurately and even for a long time beforehand known and truly foreseen, is more miraculous without divine aid, than when I take into consideration a guardian angel. The whole thing may be explained in a few words; for when the angel foresaw what was imminent in my affairs, as in the case of my son—who very likely had promised on that very evening that he would wed Brandonia di Seroni—about to marry the following day, he sent me that palpitation of the heart after a manner peculiar to himself, so that it should assume the form of a tremor of my bedroom. Then this same warning came to the servant, and thus I and my servant felt earth tremors, which none of our fellow-citizens observed, because there was no real earthquake. And had my son not married her—a thing which could not have happened without a great contention—I should not have felt mocked on this account, but rather should have deservedly prepared to render greater thanks, assuming that the angel had warned me to avoid the danger. And again, in prison, my guardian appeared to me and to my youthful associate—at God's bidding, I believe—under the guise of those violent sounds, in order that he might confirm me in my hope of divine favor whereby I should escape death, and that all those trials which I was suffering should seem less hard to bear.

I know that my angel was a spirit of power from these cases in which the manifestations were witnessed by others at the same time as myself, or in which two senses were engaged by the vision, as when the apparition said to me, *Te sin casa.* Likewise the manifestations of the year 1531—ravens, dogs, sparks of fire—came to pass because the spirit was able to work through the animate principle of beasts lacking reason, just as even men are overtaken by fear at shadows, or deluded in hope by a glittering thing like a gem or a scrap of metal.

In general, the characters of these guardian spirits among the ancients have been manifold and divers. There have been restraining spirits, as that of Socrates; admonishing as that of

Cicero, which appeared to him in death; there have been spirits instructing mortals in what was yet to come, through dreams, through the actions of the lower creatures, through fateful events; influencing us as to where we should go; luring us on; now appealing to one sense, now to several at the same time, and the more excellent for this; sometimes communicating through natural events, and finally through supernatural, which we deem the highest of manifestations. Likewise there are good and evil spirits.

Doubts arise as to why this special dispensation was granted to me and not to others, for I do not, as some think, excel in erudition, but rather the contrary. Was it my boundless love of truth and wisdom accompanied by a contempt of wealth even in my perpetual state of poverty; was it on the score of my great longing for justice, or because I would attribute all to God and claim nothing for myself that I was guided by these mystic visitations? Or perhaps my angel was present with me for an end known to himself alone.

Another question arises. Why does he not advise me openly of what he would have me know? So I might wish, but he points the way to one thing through others of another nature, as, for instance, through those confused noises, that I might have the confidence that all was clear to God though I might not see his messenger with my eyes. For the genius could warn me openly, through a dream, or through a manifestation even more evident, but the mysterious warning perhaps was more powerfully indicative of divine protection, as were those even more startling warnings which came to pass: fears, hindrances, premonitions, and ominous sounds in fearful moments. But the mystic element is necessary that we may recognize the hand of God and be taught not to forbid its working.

It is, accordingly, foolish to choose to be disquieted by things which must be revealed in this manner; it is more foolish still to wish for signs which are obstructed in so many respects by human purposes, thwarted through the commonplaceness of that to which we have long been used, and by our blind judgments. And if revelations through such channels would truly predict future events would you be able to profit thereby? And if falsely, what use are they? Yet these premonitions through the medium of my good genius are as gifts from liberal-hearted men, who are able to give many things which it is not lawful for us to receive.

Again, what is the significance of certain manifestations which could not be understood, as: *Te sin casa* and *Lamant*, and that warning concerning four years of my life from the responses of an ape? What was the meaning of the worms which suddenly appeared on the salver?[3] It is not likely that such things have happened by mistake, since a supernatural power does not partake of anything unseemly and every peculiar circumstance is known to it.

And though I have no proof of what I am about to say, it is very possible, nevertheless, that this principle—that is, the principle of this spiritual manifestation—even as nature itself, is governed through certain laws, and just as monsters are begotten in the course of nature, through no error in the process, but because of the imperfection of the material, so here the same thing comes to pass. For I do not believe that this attendant spirit is of nobler character than the intellect; it therefore errs through the fault of the medium, and on this ground must be considered an instrument. Accordingly, even as in certain years, many monstrosities are begotten because the life-giving powers of the sun have been checked, so, if the divine principle or instrument is thwarted by a certain earthly essence of the flesh, imperfections result, and a confusion of understanding of future events is conveyed through the spirit's manifestations.

If you say that those faults of nature reside in the process of action through the medium, and in the latter case in the process of action through the will, I shall answer that because the spirit is immaterial, and a good derived from God, and is indeed what the theologians name it—a good angel—it manifests truly, therefore, what, according to the will of God, is to be, nor ever errs. Nature is likewise instructed ever to indicate through the living medium the truth with which it has been endowed by the spirit, but the instrument whereby it would teach is not always well prepared for receiving this truth. And whether this instrument be but a breath of air, or something other, it thereupon assumes a form imperfect for its purpose, showing no more than the spirit wills, or than nature can, and so the error or lack of understanding arises. The only difference is that among the philosophers lack of understanding is said to arise from the gross medium not apt for receiving the imprint of truth, and among the theologians it is thwarted by sins in accordance with God's will.

Furthermore, I am not willing for many to be deceived regarding the character of my knowledge, concerning which I have spoken so often. I am aware, as it were, that I have received all things whatsoever I have known through the channels of the spirit. To what end then are my senses? Do I know everything? Then should I be as God; but my knowledge, forsooth, compared to the understanding of immortals is as the tenuous shade of man compared to heavenly mansions without bound or limit. It is, however, an understanding threefold in nature. First, there is knowledge gained by my senses through the observing of innumerable things; and this the common run of people and the ignorant are wont to magnify in me. This aspect of my knowledge assumes two questions: *What is it? Why is it?* In most cases it is sufficient to know what a thing is, because I consider it a misplaced zeal to look into the cause of all these minutiæ.

Secondly, there is an understanding of higher things obtained through the examination of their beginnings and pursued by conforming to certain principles. This aspect of knowledge is called *proof* because it is derived from the effect based upon the cause. I employ it to pass on to a wider application of the subject under consideration, or to place it in a clearer light or to give a general application from the particular. However, in this field of understanding I have less rarely arrived at comprehension by a skillful treatment than I have been aided on many occasions by spiritual insight. This form of knowledge is pleasing to the erudite, for they think it proceeds from great learning and practice, and on this account very many have judged me to be deeply devoted to study and possessed of a good memory when nothing is less true.

The third form of my knowledge is that of things intangible and immaterial, and by this I have come wholly as a result of the ministrations of my attendant spirit, through proof in its simplest aspect—that is, a simple statement of its origin, and the fact exists through the most infallible proof. However, on this point circumstances alter cases and frequently the argument leads to an absurdity. For considering the proposition that the exterior angle is equal to the sum of the two opposite interior angles—there is no reason why this should be so, but that it is so, is simply a fact.

The sort of simple proof, therefore, to which I refer is that only which deals with material forms or immaterial and is concerned therefore with natural or divine philosophy. Mathematics, however, is, as it were, its own explanation; this, although it may seem hard to accept, is nevertheless true, for the recognition that a fact is so, is the cause upon which we base the proof. Propositions also and languages—although I am not therein conversant—are not of such a nature that they come of their own accord to anyone meditating upon them.

The use of amplification and lucidity of understanding I have acquired partly from practice and partly at the inspiration of my spirit, for I devoted myself persistently to perfecting that intellectual flash of insight for more than forty years before I mastered it.

And so the whole art of writing, and of extemporaneous

lecturing has in my case been the result of spiritual inspiration and the effort I made to attain lucidity. Knowledge of this sort has, nevertheless, brought me more unpopularity than reputation among men, and more glory than profit. It has contributed rather to a pleasure by no means slight or commonplace and toward lengthening my active life; it has served me well as a source of solace in so many misfortunes, as a help in adversity and an advantage in the midst of difficulties and labors, to such an extent that it has embraced the best part of other greater branches of knowledge, and is necessary above all others for perfection and elegance.

In general, when all is said and done, the above states my case; in points wherein I can fall into error, I betake myself to my intellectual betters, that is, to the theologians.

48.

TESTIMONY OF ILLUSTRIOUS MEN CONCERNING ME

FROM FOUR enemies have come very excellent testimonies concerning me. The first came from Matteo Corti,[1] a man of the greatest reputation; when he was questioned by the Senate concerning his successor, he responded that I was the best man, and that for any duty or any requirement I was unequaled. A second was from Delfino[1] who was up for appointment as my associate in second place. He said in the presence of his pupils and openly in my presence to a certain man with whom he was talking that if I should withdraw, Montano[1] would come into the headship; and when I had interrupted, saying, "It would be rather hard for you, considering his success in winning favor with the students," he replied, "In the first place, I shall never yield my position to anyone, even if Galen himself should be lecturing. For my part I hold it more honorable if I am in second place with you than if I were elected to first place as associate of another. Nor do I deem that I should find it a more difficult matter with anyone else than with you; for with you as an associate, though I have as patrons some of the more prominent men in this whole city and in other circles of notable men, I have not been able to bring it to pass, even though exiles have been reinstated at my request, that a third part of the students attend my lectures as resort to yours."

Camuzio,[1] another candidate for appointment, had a book printed in which he seriously complained that at Pavia and in other academies my name was cited over against that of Galen; the latter ought without hesitation be put first, he said, if for no other reason than that, since he had now been dead so long, any occasion for envy would be rightly or entirely quenched and also since so many authorities were subscribing to his doctrine. This book was up for sale everywhere.

Sebastiano Giustiano, the Venetian Governor of Padua in the summer of the year 1524, was a very learned man, devoted to the cultivation of the humanities, of philosophy and theology; he had performed many diplomatic services for his country. On one occasion when he was present at a public debate in which, among others, Vincenzio Maggi of Brescia had argued—who shortly after was public lecturer on philosophy at Ferrara—he heard me, among the other contestants, taking part in the discussion; he asked who in the world I might be. They told him I was a Milanese called Girolamo Cardano. When the debate was ended he ordered me to be summoned and in the presence of the whole assembled Academy said, *"Be zealous, O young man, for you will surpass Camuzio!"* And when I, astounded by such an unheard of thing, was silent, he broke forth again: *"You have displayed understanding, O, young man; I tell you, study, for you will surpass Camuzio."*[1]

All who heard were amazed, and especially since I was not only not a subject of their kingdom, but even from a city not altogether friendly on account of the state of war which had existed between those princes and our own, long and incessantly. But since I am aware that I have many other witnesses of my reputation in writings, I have considered it worth while to add herewith the names of those in whose works I have been included with honorable mention, especially since these works, printed, are for sale everywhere.

1. Adolphus Crangius in his *Trithemius.*

2. Adrian Aleman in his treatise: *Hippocrates de ære, aquis et locis liber commentariis illustratus.*

3. Andreas Vesalius in his *Apologia Contra Puteum* which he published under the pseudonym, Gabriel, son of Zacharias.

4. André Tiraqueau, a French jurist, in the essay, *De Nobilitate,* and in his book, *De Legibus Connubialibus.*

5. Auger Ferrier in his book on syphilis.

6. The Author of the Annotations to Hermes' book, *De Revolutionibus Nativitatum.*

7. Antoine Mizaud in his treatise on Sympathy and Antipathy.

8. Amato the Portuguese in his *Commentatio in Dioscoridem.*

9. Andrea Baccio in his treatise, *De Aquis Thermarum seu de Balneis.* He mentioned me with some ill-will, as I in turn have referred to him.

10. Andrea Camuzio, whom I have mentioned above, in debate.

11. Antonio Maria, or, as he calls himself under an assumed name in his writings, *Marcantonio Maioraggio,* refers to me in his *Antiparadoxon.*

12. Adrien Turnèbe in the letter with which he prefaces his interpretation of Plutarch's *De Oraculorum Defectu,* makes derogatory mention of my name, whereby the fool indicts himself.

13. Brodeau in his *Miscellany.*

14. Buteo, a millstone who knows nothing and does not understand the art of teaching.

15. Charles de l'Ecluse in his treatise, *De Aromatibus Indiæ.*

16. The Spaniard Christoforo in his *Itinerary of the Spanish Kings.*

17. In the *Annals* of Gaspare Bugatto, where he speaks of the Physicians and Professors.

18. In the *Appendix of the Annals* of Sansovino under the Physicians and Professors cited.

19. Konrad Gesner mentions me in various places.

20. Conrad Lycosthenes: in his book on *Prodigies.*

21. Constantine: in his book against *Amatus*, especially in the place where he treats of stones.

22. Christof Clavius of Bamberg: in the third book of his *Elementa.*

23. Daniele Barbaro, Patriarch of Aquileia: in the eighth chapter of the Tenth Book of his *Commentarius in Vitruvium.*

24. Daniel Santberchius: in the seventh Book of his *Problemata Astronomica et Geometrica.*

25. Donato de Muti: in several *Aphorisms.*

26. *Epitome Bibliothecæ.*

27. Francesco Alessandrino: in his *Antidotarium.*

28. Francois de Foix, Count of Candale, whose censure I took for praise: in his *Geometria.*

29. Francesco Vicomercati in his *Commentarius in Libros Meteorologicorum Aristotelis.*

30. Fuchs in his medical *Compendium.*

31. Caspar Peucer in his *Commentarius de Præcipuis Divinationum Generibus.*

32. Gaudenzio Merula, a native of Novara, in his article *De Bello Erasmicano*; he was the first who ever mentioned my name in a printed work.

33. George Pictor, a physician, mentions me in his books; of this more later.

34. Guglielmo Grattarolo of Bergamo, a physician.

35. Gabriello Fallopio: in his *Liber de Metallis et Fossilibus* in which he freely contradicts me.

36. Guillaume Rondelet: in his *Historia Aquatilium.* He mentions my name in derogatory manner.

37. Gemma Frisius: in his *Arithmetic.*

38. Girolamo Castiglioni in his *Speech in Praise of the Fatherland.*

39. Hieronymus Tragus: in his *Book of Plants.*

40. Jerome Monteux, personal physician to the French king, referred to me in his writings.

41. Jacques Peletier: in his *Mathematical Works.*

42. Jean Duchoul: in his *Historia Quercus.*

43. Juan de Collado: in his Book, *De Ossibus.*

44. Giambattista Plozio: in his *Tractatus de in Litem Jurando.*

45. Johann Schöner: in his treatise, *De Nativitatibus.*

46. Johann Cochläus: in the introduction of his Historia.

47. Joachimus Scelerus: in his introduction to the *Works of Juan of Seville.*

48. Johannes Ceredus: in his work, *De Aquarum Elevatione.*

49. Johannes Stadius: in his *Tabulæ* and *Ephemerides.*

50. The Portuguese, João de Barros: in the fourth Chapter of the first Decade of his *History of the East Indies.*

51. Julius Cæsar Scaliger: in his treatise, *Exotericarum Exercitationum lib.* xv *de Subtilitate ad Hieronymum Cardanum.*

52. Jacques Charpentier: in his *Epistola in Alcinoum.*

53. Ingrassias: in his book, *De Tumoribus.*

54. *Liber Aggregatus de Aquis.*

55. *Liber Aggregatus Primus de Gallica seu Indica Lue.*

56. Livin Lemmens: in his book, *De Naturae Secretis.*

57. Lorenzo Damiata: in his not yet published *Geography.*

58. Leo Suavius: in his work, *De Arsenico et Auripygmento.*

59. Luca Gaurico: in his *Book of Nativities*; his reference to me is unfavorable.

60. Mattheus Abel: in his book, *De Situ Orbis.*

61. Martinus Henricus: in his *Quaestiones Medicae* and indeed in various places.

62. Melanchthon: in the Introduction to his treatise, *Doctrinae Physicae Elementa.*

63. Melchior Wieland, a Prussian.

64. Michael Seifelius: in his *Arithmetic.*

65. Michele Bombello: in his *Algebra.*

66. Niccolò Tartaglia, who spoke evil of me and later, in Milan, was obliged to take it back publicly.

67. Philandrier: in his *Annotationes in Vitruvium.*

68. Pierre Pena and Matthias de Lobel: in their work. *Stirpium Adversaria Nova,* in the chapter, *De Hora et Antithora.*

69. Reiner Solenander: in his treatise on warm waters.

70. Severinus Bebelius: in the Second Book of his work called *De Succino.*

71. Taddeo Duno: in his most important work.

72. Valentin Naibod of Cologne, in *Commentarius in Alchabitium.*

73. Vareus in his *Materna Poesis.*

I know that many others, whose names I do not now recall, have made my name known in their writings. Of those who have spoken ill of me I am not aware that one has gone beyond the elements of grammar, and cannot understand by what impertinence they have managed to get themselves into the ranks of the learned. I refer to the following:

Brodeau
Fuchs
Charpentier
Turnèbe

Rondelet
Buteo
De Foix and
Tartaglia

Scaliger, Duno, Ingrassias, Gaurico, and Solenander contradicted me for the sake of making a reputation for themselves.

Observe now other testimonials of men; for no one doubts that as regards commendatory reference in writing, neither to Galen nor perhaps to Aristotle while they lived was this accorded nearly as often as to me, acknowledging of course that it was my lot to profit by the art of printing. Andrea Alciati was the first, as I have elsewhere related, and whom I name, *honoris causa,* in addition to the fact that he was wont to call me the *Man of Discoveries*, and daily he turned the pages of my books, especially those which have been written on Consolation. Ambrogio Cavenaga, the Emperor's personal physician, used to call me the *Man of Many Labors.* Julius Cæsar Scaliger ascribed more titles to me than I should have thought of arrogating to myself, calling me *ingenium profundissimum, felicissimum et incomparabile.*[74] Nor was it possible for me to be treated so invidiously that my name be ignored in the classrooms of Bologna, Pavia and elsewhere. Angelo Candiano and Bartolomeo Urbino, both illustrious men and famous doctors—O, how often have they been found, and they made no attempt to conceal it, with my books in their hands although they were not exactly my friends!

But let us put an end to this, lest we seem to follow the wraith of a dream, for all mortal things are futile and praise of them is altogether empty.

49.

MY OPINION CONCERNING WORLDLY THINGS

THERE ARE two sources of very great unhappiness in mortals; one is, since all things are empty vanities, man seeks something which is ample and substantial. Who indeed does not think he lacks that tangible good? The sick man feels the lack of sound health; the poor man, of resources; the lonely man, of children, and the unhappy man, of friends. And while each seeks and finds not, he is tormented; when he at last finds, more and much more, he realizes that he has been deceived, for there is always something wanting; did not even Augustus complain that he lacked friends and bemoan the want of decency among his descendants? These mortals deceive themselves.

Another source of unhappiness those experience who think they know what in truth they know not; such are also deceived and fool others as well. Still another group make pretenses and thereby fool their fellows.

To the foregoing causes of misfortune two other accidents are to be added: to live in a place and at a time when the government is wrecked through revolution is unfortunate; for to have the will to resist at such periods is for the individual to encounter innumerable hardships, to suffer many anxious moments, and is, furthermore, folly; yet to sidestep the issue is none the less dangerous or foolish. For property and money are at the mercy of public calamities. A second accident is the constant ebb and flow on which the numbered days of life are borne, so that many in their own struggle for life, and for the sake of another's profit, perish.

These things are hardships for all, and for old men, and for the incautious, they represent the greatest difficulties; to those untried or not on their guard they are insurmountable; and they are made worse through the effects of another's folly which, when coupled with ignorance of the outcome, renders evil men even baser, both of themselves and as regards others.

Some, urged by misfortune, join themselves to protective societies; to others the idea of God and death must be presented: God as one who is unerring; death as the last of evils, which, certain and about to put an end to evils, forces us to despise other unhappy conditions. Again, it is good that you should support your happiness by many interests lest, having but one which fails, either you fall, or you become the slave of your own misfortunes. Thirdly, you should not judge things by their quantity but by their quality, for humble beginnings of significant events ought to be preferred to imposing beginnings of lesser character. And though one may lack many things, and since all things may not be brought under man's foot, we ought not to devote ourselves intemperately to any one pursuit, except in the measure which necessity or security may require. Therefore not equally, but proportionately, should we exert ourselves to acquire virtue, which should not be lacking in any, and property, which is a source of help to all.

A sixth consideration is that training can do very much for children; but if they do not profit by their instructions, and they are evil by their very natures, or foolish, or insolent in their liberty, and, even after they have left their childhood behind them, are still mean-spirited, though a man have but one, and be burdened with years besides, this is the quintessence of all evils, counting even poverty, lawsuits, loss of patrimony and hard times, all of which have been my lot in addition to this last misfortune.

Finally I declare that by the side of all that might still be said, there are actually some solaces at hand: one, you should reflect that if you possessed nothing you would lack still more; another, you should make haste to find a Scipio from among your friends, to find a daughter-in-law or a relative; in this it behooves you not to err, for therefrom you may with all your might, as if reborn after your follies, bend to your tasks anew.[1]

Well, then, to come to a brief conclusion, since all things are insignificant and vain, whatever indeed concerns our remaining activities depends thereafter upon even the most fleeting circumstances. In my own affairs particularly, by this one event, among many other examples, the foregoing truth is made plain.

In the year 1562, on October 17th, if I am not mistaken, I was in Milan, and about to set out for Bologna. About six days earlier the brass tip had fallen from the garter by which I bound my hose to my doublet. I had neglected it because I was pressed for time with much business. However, there were six sets of garters, purchased the day before the clasp had fallen off the old garter, which I had packed with the intention of taking with me to Bologna. The day and hour for my departure had come, when as usual, in the very instant I was about to mount the carriage, I felt a desire to relieve myself. This done, while trying to fasten the garter I kept fumbling until, thoroughly annoyed at the inconvenience, I went out to the shops which were around my house—there were three, I believe—to buy a garter, but I found none anywhere. Hesitating, then, as to what I should do, I remembered the sets of garters I had formerly purchased. I sent for the key from my son-in-law who returned it. I opened my chest, which, being of German construction, was unlocked with some difficulty. There I saw the sets of garters, and there, behold! my eyes fell upon the stacks of all the books I had written, and which I had laid away in the chest against my coming journey, so that I might transport them all with me.

Obstipui, steteruntque comæ et vox faucibus hæsit.[2]

I took them all up and carried them along with me. I began my lectures. About the first of December letters came saying the chest had been broken open by night, and whatever was within had been stolen. If it had not been for my garter I should not have been able to give my lectures, I should have lost my post and gone begging, so many memoirs would have perished, and I should have died ere long of grief. And all this depended on an insignificant instant! Alas for the unhappy condition of mankind!

To possess a truly discerning understanding of the *mores* of men you will look into their cultivation of instructive disciplines, and examine the degree of perfection attained through native character, law and custom. Uncultivated men are, accordingly, simple and indomitable; and consequently given to extremes. When they are good, they are the best of men, for they are not corrupted; when they are bad, they are the most evil of men, for they are bound by no reason nor moved by any persuasion: as to their lusts, abominable, and unseemly in their gluttony. They are bitter in their wrath, especially the poor, because of their greed, and the rich, who are ambition-mad.

The slothful are both crude and invidious, and as such, evil-minded and even greedy. Those of them who live under a tyrannical government, if they have influence, use it to rob others of their property; if they are poor they cling tenaciously to what is their own. Both circumstances beget avarice; there is neither love nor trust nor mercy. Boldness, born of their bitterness, makes them cruel men; their slothful lives make them sordid, particularly when their idleness is accompanied by gluttony and lust. They are given to industry when necessity or social pressure demand, and make a display of cleverness when urged out of the jejuneness of their existence. If by law they are prohibited from instruction, and honors accrue to their handiwork, and ambition has a place—as when they are ruled by the aristocrats—they distinguish themselves through their devotion to the arts and crafts, especially if their region is one thriving by reason of a variety of industries. In a republic, or where wealth alone is sought, there is little regard for honor one way or another.

I have given an example of the irony of human affairs; now I shall add an incident the like of which happened on innumerable occasions; you will laugh, and yet on no more than this, death or life depends: it happened only yesterday, that is April 28th, 1576. Since I wished to make my way to a jeweler through a narrow alley, I ordered the driver, a dolt of a fellow, to go on to the Campo Altovitaro. He said he would go, but understood me to say another place, whither he went. I returned to the place appointed after my transaction and found him not. I suspected he might have gone to the plaza of the warder of the castle. Thither I trudged, loaded down with cloaks—put on for warmth while driving—and in the course of my walk I met Vincenzio, a Bolognese and a musician who was my friend. He observed that I was without my carriage. I went on but the carriage was not to be found. Suddenly the greatest anxiety swept over me, for I had to retrace my steps over the bridge although I was exhausted with weariness, hunger, and profuse perspiration. I might have asked for a carriage from the Lord of the Castle, but certain dangers involved in such a request restrained me from such a course. I commended myself to God and considering that wisdom and patience were necessary I returned, but firmly minded neither to give out nor to take rest. Yet at the end of the bridge I betook myself to the Altovito Bank where, under the pretext of getting some information I desired about rates of exchange on Neapolitan money, I sat down. The agent was cheerfully making explanations when the Lord of the Castle entered, so I withdrew immediately, and there in the middle of the plaza I saw my carriage. The driver had been advised by my friend whom I had met; I mounted the carriage, doubtful whether to go on my way because I was so hungry, when in my pouch I found three raisin clusters. And so I finished my round of calls safely and even with pleasure. Here you can see how many momentary sequences befell: first I met Vincenzio, then he and my driver met, followed by my plan of going into the money-changer's and finding him free from pressing duties; then the Lord-governor came in and I went out, whereupon I ran across my driver and later discovered those raisins. There were seven incidents, a single one of which, had it happened either sooner or later by so much as the space required to utter two words, would either have been the end of me or brought me the greatest discomfort and annoyance. I do not deny that to others also such things occasionally happen, but rarely incidents that so hang upon the instant, and are so fraught with risks and involved in greatest difficulties of which they themselves are unaware.

50.

FAMILIAR SAYINGS

THOSE THINGS should be done the recollection of which it will never at any time be a matter of regret to summon; things, the aptitude for which desert a man sooner than the will, and things in the memory of which there is peace of mind; and better it is, even, if that memory seek to avoid any solicitude.

When there is a choice between contrary courses, that ought to be held for best which in the long run brings the least evil.

As Sorano used to say, it is honorable to accept from those bringing you offerings, but not to make exactions from those unwilling to give.

In a doubtful choice, select what seems naturally the more favorable—the course which makes some show of good return; for if you choose that which is merely your preference, and no opportunity to retract is given, you are confounded by a questionable choice, and forfeit prestige as well.

He who is unwilling to listen to reason is a brute; he is therefore either worthy of blows, or you must part company with him. Man alone conducts his affairs and looks out for his interest by talking them over.

Men morose, obstinate, harsh, and sluggish are useless on the principal count, as in many other points as well, and therefore to be avoided above all others.

To the powerful especially, either of the following rebukes is sufficient, and neither forfeits the modesty a man should maintain: *You have inflicted an injustice upon me,* or: *I am complaining because of an injustice done me.* To those nearly related, as to heirs: *After my death will it be your wish that my property had not been carefully protected?* You will confound them thus.

To one who was twitting me with the small number of my pupils I responded: *More copies of Donatus'*[1] *works are sold than Vergil's.* And, criticized because I stood alone in my opinions, I said: *It is on this very count that the unicorn is worth so much.*

To a jurisconsult also twitting me about the paucity of my students I responded: *Some students confer more distinction by their presence than others dishonor by their withdrawal.*

To a doctor boasting that he had more patients than I, I replied: *That is not the question, but rather how many of these are healed.* To others I retorted more sharply: *It is a disgrace for so many to perish in your hands.*

Exhorting a young man to shun the society of worthless men I said I would show him the apple which corrupted the heap but I could not show him the heap which would restore the rotten apple.

To those upbraiding me because I brought up so many boys I replied: *My deserts in this are double: I do good, and hear evil of it.*

Wisdom, as many another priceless thing, must be dug from the very bowels of the earth.

When a man was comparing me to other scholars I quoted this Vergilian verse:

Quia si idem certet Phoebum superare canendo?[2]

I have frequently bidden those about me to meditate on this especially, this alone: *What is the greater, what the less?*

It is greatly indicative of prudence to have a distinguished friend.

To a man who reproached me on the score of my old age I said: *That man alone is old whom God has deserted.*

Friends are your support in adversity, flatterers bring you advice.

Evil must be healed by good, not by evil.

I know that the souls of men are immortal; the manner I know not.

I owe more to bad doctors because they lost me my enemies, than to the good ones, although they were my friends.

When some evil-minded men taunted me with having erred in a diagnosis while I was in the company of other evil and ignorant men I said that it had been a marvel if in such a company I should make a true prediction or indeed do any other thing rightly!

When you are about to do something, think ahead and see what your status will be when you will have done it—whether it will turn out as you hope or not.

An illustrious man ought to live under the ægis of a prince.

Receive your friends with a joyful countenance, as they have deserved that favor, and your enemies also, that you may prove the better man.

Those who write uncouth verses are like those who devour raw food: for a slight pleasure subjecting themselves to serious troubles.

You may measure the trustworthiness of men by what is expedient for them unless they be in every respect more magnanimous than faith itself.

The greatest thing in human affairs is to determine the goal of one's endeavors.

The most cultivated men of our epoch, because they cleave only to material things of immediate import, are condemned in one breath of impiety, ingratitude and ignorance.

Likewise, as I was about to dismiss a servant, I said: *You give me satisfaction, but I do not please you, and on this account you force me to abandon you.*

When someone asked me: *Why, when you are so wise, are your sons so foolish?* I retorted, *Because I am not so wise as they are foolish.*

Those who are well favored by the gifts of fortune are like

small boys leaping down from the steps of a stairway: the more steps there are the happier they are, but they vie with greater danger to themselves and with greater loss.

It is better to leave a hundred things unsaid than to utter one which should have been passed over in silence.

School boys should be obliged to give ready responses, to the end that they may concentrate, and not simply for the purpose of giving an answer, in passing, to the questions asked.

To somone asking what went on at Rome I said: *What goes on in the Queen of cities and of the affairs of men!*

And have you been in prison? I was asked. To one I said, *And do you wish that you had been there also?* To another, still more weak-minded, *What have you done for which you fear arrest?*

Those themes should not be introduced into books which are not pertinent to the end or not worth reading.

When you say anything artfully malicious you ought to have a thick skin; when, then, you hear anything of the kind, consider that the speaker has a thick skin.

In matters of business, in contrast to matters of theory, a certain acquaintance with the subject is of some help, but it is necessary to have quantitative knowledge. One can do good by offering rhubarb to a patient sick with tertian fever even if one does not know the exact quantity, but it is better to keep silent or not to visit a patient if one has not become acquainted with the manner of treatment to be employed.

Tears are the medicine of grief, indignation of pity, and the inevitableness of both has been amply attested by history.

In general, in all things it is good if one can sue for time.

In your negotiations confront an offense with offensive measures; meet negligence with a lawsuit and exactions; meet obstinacy with irascibility, arrogance with harsh treatment and scanty fare, and those who deal with fists in place of words resist by making as if to send a hasty message to the authorities.

When you wish to wash yourself first prepare a linen towel for wiping.

When about to employ an old woman as a housemaid, ask whether she knows how to sew, wash, and bake bread; order her to hurry around and build a fire. You should complain also that some of your wine is lacking. Ask if she have any relatives or friends as if you had need of these; inquire under what circumstances she left her former master, how many husbands and children she has had. Then leave her to her devices, or at least take care to set a suitable watch upon her.

To be quick to tell all you know is the worst of faults, yet in the course of events sometimes necessary; in deliberation there is need for moderation.

Do not make demands upon another's rights, above all from the lords of the earth; whatever is within your own right do not demand on condition, but use your judgment, nevertheless, with modesty.

With men you may not say all you feel, but look well to whom you give much more than a loan.

Time gives us but a shadowy image of the significance of God in eternity, and even this is imperfect on account of constant change.

In perilous circumstances or those exposed to malicious report either as regards your affairs or yourself, if you be not certain that you are able to speak conclusively, it is better to let the matter pass; in this respect many are at fault since they are too ambitious for the reputation of knowing much or of having always been successful.

Take care lest you give over your interests into the hands of one who enjoys your favor: will he make use of your property if you leave it with him? And if you would take it again, a danger arises that he may drag you into an odious relationship with himself, and he will surely hold you under obligation.

See that a book fulfills its purpose, and that the purpose rounds out the book: such a book, and no other, is perfect.

To one who said to me, *I pity you*, I replied, *Without good right.*

Evil is but a lack of good, and good is of itself a virtue which is within our power to possess, or rather which is indispensable.

If you have not riches, children, and friends, yet have other possessions, you are fortunate; but if all other possessions are lacking, you are due to endure but a brief space.

Although there are many arts, one is the master of all: that fundamental precision by means of which one may explain many things through a few cases, or render vague facts clear, or state in certain form facts formerly doubtful. However, there are three requirements: that all these general principles fall in with that one master theory, that they harmonize one with the other nicely, mutually inclusive and exclusive, and that each may be peculiarly adapted to its own special usage. Aristotle passes over this last alone on account of the limitations of science in those days. Remember also that it is legitimate to concede something to grace of style, as in the writing of stories.

Many wrongfully complain about virtue saying it is in slavery to fortune; others deem it to be the mistress of happiness—words worthy in truth of a high-spirited man.

> Victrix fortunæ sapientia, ducimus autem:
> Hos quoque felices, qui ferre incommoda vitæ
> Nec iactare iugum didicere . . .[3]

Yet he errs in double fold: first by teaching that understanding (meaning our human understanding) is more potent than fortune, although our everyday experience is to the contrary. The reason is that fortune displays her whole aspect and lets loose all her powers in any circumstance whatsoever, and we have only a small offshoot, very frail and slender, of understanding. Fate is not, therefore, more powerful than understanding, yet by much less is our human understanding the Victor over Fate. Yet Fate yields to divine wisdom nor dares to set foot where she has perceived the very odor of Wisdom passing by. The expression of Brutus as he lay dying likewise hardly pleases:

> Infelix virtus, et solis provida verbis:
> Fortunam in rebus cur sequeris dominam[4]

even though Plutarch brings forward Antony's testimony about Brutus, that he alone slew Cæsar because of the glory of the deed, the others because of envy. Perhaps this came about, however, as the result of other motives than any he expressed. For the following appears as Cicero's testimony in his letters to Atticus: *Brutus had resented bitterly many things which had happened and was grievously troubled about events which succeeded*, a circumstance which would never have existed had the end of his action been glory!

Unjustly, furthermore, Cicero complains because Antony, who had spent his whole life in the pursuit of arms, was preferred by Fortune to himself whose existence was more sheltered and actively devoted to oratory.

The terms men use are stout, rude, and bristling: their meaning, however, is none the less vigorous; compare the references to Nero, who is the subject of the writings of Tacitus. How foolish, then, for Brutus to seek a place for noble virtue in the midst of his sedition. Since in a city besieged by an enemy host a happy man cannot be found, how much less in a city rife with revolt? And so happiness is not found in Fortune but in Goodness. Fortune, nevertheless, is able to offer more obstructions than Virtue can array herself against when Fortune is her adversary.

Three things above all others change their ways: age, fortune, and married life—therefore take care. Associations with one's fellows also change, and even as glowing iron, nothing touched by human hands is worse; yet that very iron beaten by hammer becomes a source of gain for the smith, and a useful article for others.

A LAMENT ON THE DEATH OF MY SON

Who has snatched thee away from me—
O, my son, my sweetest son?
Who had the power to bring to my age
Sorrows more than I can count?
Wrath in whose soul or what stern fate
Willed to reap thy youth's fair flower?
Not Calliope, not Apollo,
Served thee in thine hour of need!
Cithara, now, and all song be still;
Measures of threnodies shall renew
Mourning and sighs for my dear son.
—Strains of his singing haunt me still—
Laurels, alas, in the healing art,
Knowledge of things, and a facile gift
Of Latin tongue—what profit these
Labors long if they swiftly die?
Service rendered Spanish prince,
Duty done to the noblest of men
Help thee naught if with these for thy judge
Death with his scythe doth seek thy blood.

What, ah me, shall I do? My soul
Swoons to remember thee, gentle son;
Silent, I brood on thy destiny grim;
Tears that I dare not give to words,
Shall I not shed for my stricken son?
Lasting encomium had I reserved,
Fitting reward to thine ashes paid;

Silence—O shame—must my tongue now guard,
Death unjust nor its cause announce.
Grave are the ills thou hast borne, mild son.
Prince and Senate and ancient law
Ordered thy doom whilst thou in rash haste,
Brought an adultress the wage of her crime.
Safely adultery now in our homes
Mocks and insults when punishment swift
Stays the avenging right hand of the youth.

Son—the reflection true of the good
Strong in my father—worthy to live
Long through the years—Alas, my beloved!
Fates have forbidden and swept all that good
Far past the stars, and removed from gray earth
Every bright and illustrious thing.
Hail thee, child, for thy spirit high!
Clear is thy blood from ignoble stain;
Honor of forefather's hast thou sought.
Far stands the king, and hope of safety,
Phœbus denies the lands his beams,
Light from Diana passes and dies,
Stars in the calm sky glance no more
Lest they look down on a palace foul,
Stained with the reeking blood of the slain.

Where lies my way? What land now claims
Body and limbs disfigured by death?
Son, is there naught but this to return?
Thee have I followed on sea and on land!
Fix me—if mercy is anywhere found—
Pierce me with weapons, O ye mad Gods!
Take with thy first blow my dreary life.
Pity me thou, oh great father of Gods,
Thrust with thy spear my hated head

Deep into Tartara; else am I bound
Hardly to burst this life's bitter chains.
This, O my son, was not pledged to thy sire,
Love so unholy to trust with thinc all—
Love that has ruined thee, son of my heart!

Wife of a memory blessed and true,
Happy thy death, nor spared for this grief!
I, through this crime, have myself brought disgrace,
O son to our name, for by envy compelled,
Homeland and Lares paternal I left.
Death had I sought for my innocent soul,
But surviving and living I vanquished my fate.

Ages to come will know, son, thy name,
Orient lands will hear of thy fame;
Dead to us thou art indeed—
Life hast thou won through all the earth!

For all of us alike must die, and, as Horace says, naught remains but a glory which waits for its splendor upon the herald of our virtues, a bard, inspired and therefore famed afar; to this end Alexander kept longing for a Homer. History, moreover, diminishes glory by the illustrious deeds of succeeding generations. The poets relate their tales and are, accordingly but lightly esteemed therefor. The great marvel is the case of Horace who without writing an epic dared to maintain the perpetuation of his fame, since without historical poems men could not be entertained; of this Horace himself gives testimony:

Lectorem delectando, pariterque monendo.[5]

But what would have been an obstacle to another—the swift decline of the purity of the Latin tongue—redounded to his popularity because his own purity of diction was so exceptional.

The achievement of a glorious and celebrated deed should abide; the story should be composed and magnified with the imagery of poetry by the bard, because a bare narration might dispel its unity of atmosphere. Whence:

> Graiis dedit ore rotundo
> Musa loqui.[6]

The poet, furthermore, ought to be inspired to his task of adorning his writings, and every principle of composition should be observed which Aristotle strove to embrace within the limits of his trustworthy treatise.[7] When all these four precepts are observed at the same time a man may hope for an eternity of fame; otherwise not.

But why all this discussion?

*H.** In order that we may know what conditions are necessary for a happy life; and, since there is no true happiness in this existence, to keep us from rushing on in vain search of it, whereat we may be burdened with unhappiness.

*S.** Yet to tell the truth these philosophies are not enough. For it is as well our duty, as that of all other artificers, to know the final purpose of all we seek. The smith knows and teaches how to make nails, bars, anvils and sledges: the nails for the purpose of fastening together boards, the bars for bolting doors, the anvil for sustaining the blows of the sledges. "But you," say the artisans, "give no instruction like these

---

*H. = Hieronymus; S. = Perhaps "a Scipio" whom Cardano chooses for his friend and interlocutor.

practical teachings. It is not clear what happiness is, nor do you admit that it actually is. Nor do you state what the function of happiness may be; this is plainly very difficult because we do not know, with this inanity and vacuity of yours, whether you have, by your doctrine, advanced our understanding, or whether there be anything at all in your vain philosophy." Wherefore they wish to hear what utility, if any, there is in your theory, or what purpose. For if there be no advantage, why write, teach, learn?

H. There is a purpose, and there is not. First, this: we know, as I have said, that the sum of unhappiness is greater than happiness; the latter scarcely exists. And in all this emptiness there is sometimes some infinitesimal and fleeting good. All, nevertheless, amounts to nothing except as we have learned to appropriate at the right moment this brief good and to avoid misfortune; learned to thrust away and reduce to naught anything calamitous—provided only it be not the extremity of misfortune, by men's comparisons, for such is possible; provided, I repeat, the evil be tolerable or even moderate, but by no means the most appalling of adversities; whereas, in ignorance of these attitudes your lot becomes wretched indeed.

S. Therefore, as I see it, the advantage is fivefold. First, you may alleviate misfortune, provided it be not the extremity. Second, that you may augment that moiety of happiness which is humankind's portion. Third, that you may appropriate some infinitesimal blessings, and in a short time some degree of abundance will be found in the emptiness of your existence. Fourth, that we are to know there actually is some very small, ephemeral happiness, and that virtue is the only source of it, and it is not thwarted by even great detriments.

Fifth, that such good fortune can be somewhat prolonged because the underlying causes of it and their functions can be extended in certain cases quite appreciably, especially in view of and in comparison with the brevity of human existence. But if the span of life should be five or six hundred years, all would end their lives in desperation; and this tenuous happiness would likewise vanish.

H. Well done! And I thank you because you have set forth my propositions much better than I should have known how to do it. I add simply this, notwithstanding, that this same good fortune, howsoever entirely insignificant and slender it may be—indeed, to be frank, next to nothing at all—has yet four grades. First, when it is in the very act of coming to pass, a momentary condition, it seems to be something, but it is not. Proof of this is the consensus of the philosophers who call the sensation nothing but freedom from pain.[8] The second stage is when it has just passed over and the causes and effects of it still remain, so that it actually seems present. The third stage is when the time has passed yet the memory is still fresh, though shadowy and without substance. At the fourth stage no trace of all these states remains; they have become like to the casual events of every day, the memory of which has passed away; although we may still recollect, the recollection is without any effect. All of which goes to prove that it is enough to have traversed the span of mortal life without experiencing a calamity of any magnitude.

51.

THINGS IN WHICH I FEEL I HAVE FAILED

LIKE THE Trojans, who learned too late their folly,[1] some of us are slow to realize that it is impossible not to make mistakes. But it is inevitable that those err many a time who wish to be in service to pleasure. My most serious error was in the education of my sons, for a proper rearing can accomplish the most far-reaching results. But I lacked means, prudent sons, brothers, sisters, relatives and friends, resources, influences and faithful servants. I was able, however, to choose, had I been willing, to forego much of my writing, forcibly to curb my disposition, and to refrain from the pleasures of the flesh; and, had I wished, to be more assiduous in cultivating friendships, and to await the decree of the Senate of the University at Bologna; for, even as there was some advantage in not having solicited the post, how much more in not pressing my demands? A favorable occasion is of more avail in situations of this sort, when rivals are opposing you and the power of your friends falls short of your needs.

My devotion to the game of chess was really not much of a vice; yet I cleave more to that sentiment of Horace:

> Vixi; cras vel atra
> Nube polum, Pater occupato[2]

with this additional advice: let the past not stand in the way of your future welfare.

Besides these shortcomings I applied less attention and zeal than I should toward restraint in the matter of food and toward acquiring some grace in the art of conversation. It was, therefore, not enough to have observed the seven axioms in my conduct as a whole, and what was the good? Although I gave myself over intemperately to the study of letters, the same steadfastness in the rest of my interests would have resulted in a less profitable devotion to literature. It was far in excess of enough to have lacked so many things—a good memory, a knowledge of the Latin tongue in my youth, unimpaired health, friends, a family prepared to offer some advantages, virility throughout a long period, a certain graciousness and winning manner, sons endowed with at least common sense. As against these, of timid heart I had a superfluity, and of endless lawsuits, morose-spirited elders, and years turbulent with wars and heresies.

But, you may say, are you not able to help your condition by means of the things you have learned? The truth ought to be prejudicial to none: I have found that in large measure the things which others have discovered are erroneous, or else I have not known wherein even an unprofitable employment of the majority consisted. My own discoveries are more useful, but they came too late; had it been permitted to come in good time upon facts which I later had in my possession, or if the things sought by others had been recognized by me earlier, perchance I should not have found life so difficult.

Nevertheless I still have so many blessings, that if they were another's he would count himself lucky. I have the knowledge of many sciences; I have an heir of my line, of unblemished name, though shadowed by misfortune; I have books published, and many ready to be published; I have reputation, position, and substance honestly acquired; powerful friends; an understanding of many mysteries; and, what is best, a deep veneration for God.

Yet, as I have so often said, the achievement of all things cannot fall to the lot of one man, nor can he excel in every field; nor can he, although he greatly excel in one particular only, even in this gain perfection. What, therefore? Would you require in me that which is denied to every class of men? Would you wonder that I had made my mistakes since all other men miss the mark?

52.
CHANGE FROM AGE TO AGE

THE SUCCESSIVE periods of our lives bring about changes in our customs, in the bodily form, in temperament and in appearance. I have heard it said that in childhood I was fat and rosy; in boyhood I was scrawny, with a long face, fair complexioned and flushed, and I grew so rapidly that by my sixteenth year I had almost completed my growth, and I seemed as tall as I am now; my disposition was inclined to sadness. In youth I was sandy haired, in deportment and temper nothing unusual, cheerful, given to pleasures, especially to music.

In the period of my life falling between my thirtieth and fortieth years there was little change. I had troubles which disturbed me—poverty, with a wife and children, an infirm state of bodily health, and rivals so bitter that when I cured Bartolommea Cribella, a noble lady, and following her, her brother, this same man, as he was convalescing, made mock of me. To him I said, "*What would you all be doing*"—for others had joined his mockery—"*if he had not been cured?*"

Was it not then to be expected that after all this—but not before my thirty-ninth year—I began to get my breath?[1] Nevertheless, during the next four years, that is, from the first of September 1539 to the first of November 1543, I did nothing else except strive, by private means and public measures, to the end that I might be freed from my embarrassment and given the honor due me as a doctor. The first year of changed fortune was, accordingly, my forty-third. From that time up to my seventieth year almost twenty-seven years intervened—

the duration of the Peloponnesian War—in which I composed all those books. From my seventy-first to my seventy-fifth year, four years have glided by during which I wrote twelve works, consisting of eighteen books. But the greater bulk of my literary labor will be that of my published writings consisting partly of old teachings, partly of new.

During those early years I devoted seven years to pleasures—music and others—to gambling and to fishing, the latter especially. Later I trained myself for debate, and at the same time found myself still more lacking in bodily health. My teeth ached and a few of them fell out. Gout invaded me, but did not, however, torture me. The attack was generally of twenty-four hours' duration, and whatever pain remained tended toward a gradual moderation. Up to my sixtieth year my vitality did not diminish, so that it appears my strength failed rather because of mental anguish than because of my age. From that year on I have made it my business to attend to my domestic affairs, but so many discouragements have cast their baleful light upon me that it is a marvel that I have been able to live until the present. If anyone should enumerate my struggles, my anxieties, my bitterness and grief, my mistakes in conducting my life, the claims of my personal interests, fear of poverty, sleeplessness, intestinal trouble, asthma, skin disease and even phtheiriasis, the inconstant character of my grandson, the sins of my son, would that man not be surprised that I still survive? Very many of my teeth have fallen out, so that but fifteen remain, and they are neither whole nor solid.

So many the plots against me, so many the tricks to ensnare; the thieving of my old housemaids, drunken coachmen, and all the lying, cowardly, faithless, arrogant crew that it has been my lot to deal with! I have had no one I could rely upon, not one who was half reliable. Six times, or thereabouts, I have firmly believed as I lay down to sleep that I should nevermore live to wake again because of my many burdens and the error of my ways. Twice, tormented by these, I thought I should die in the night. I have not yet drawn up the will which I wish to be my last.

How, you will say, was it given you to struggle clear of so much trouble? Deep grief was grief's own medicine. I confronted disdain with righteous anger, and devoted myself to serious study as a counterpoise to an insane love for my children. My petty sorrows I dispelled by playing at chess, while to great afflictions I offered a show of false hopes and plans. I did not breakfast, but my appetite was so reduced that in the morning I was satisfied with a baked apple or with fifteen dried Cretan grapes, without wine, often without water, or at least very little of one or the other. Lately I have adopted a taste which is pleasing and I hope wholesome—the "white broth" of Galen with bread simply dipped therein, and nothing else besides. My supper is somewhat richer.

In all vicissitudes, there runs through my mind recollections from my book *On the Best Way of Life.* I keep thinking what happened to the son of Sulla whom Cæsar ordered to be slain with his wife because she was Pompey's daughter.[2] And what to Quintus Cicero? What to his brother Marcus? His daughter died without an heir even while the father still lived, and for this he lost his reason. And although he had a surviving son, it was as though he had neither son nor daughter. And I remember unhappy Terentia, living on when all her loved ones had perished, with the memories of her hundred years,[3] and of her once so flowering married life. O, the irony of mortal vicissitude! What has become of the works of Theophrastus,[4] so lovely and so useful?

After long experiment I have settled on a supper of good-sized fish, tender and properly prepared. To provide the most nourishment let them be gently boiled together and the firm parts be used for food. I usually select a carp, but here in Rome they are not to be had. In lieu of this I eat turbot, flatfish or a pike weighing from three-fourths to a pound and a half; also any tender, wide-bodied fish, or even mullet. For reasons I have elsewhere given I avoid pond carp—they call them *Scardeas*—but not brook carp. A good dish is broth of beets with garlic, or of cockles or crab or snails made with green bay leaves. In place of beets or sow-thistle or endive roots, fresh leaves of borage and radishes do for the salad course. Or in place of one of these, the yolks of eggs—fresh—with one of which I am often contented. I eat white meat, separated and well roasted, and calves' feet, the liver of fowls and of pigeons, as well as the brain and any part of the giblets. Meat is good browned on the spit, then cut into fine or thin slices and beaten until tender with the back of the butcher-knife, after which it should again be turned over and over for some time in hot goat's suet. I prefer to all other meat dishes, a pot roast in its own broth.

When my feet are cold I wash them, for thus their warmth is restored again. Neither do I eat my food unless it is warmed and the dry parts moist. After breakfast I do not walk about, much less after supper. I become more tranquil in spirit and more courageous in proportion to my understanding and not by virtue of this regime. And since death seems to me rightly to be feared, therefore I hate it. Let us leave it to others who do not shudder at its coming, to run to meet death, after the fashion of Taurea of Campania,[5] if to await it is not enough.

But I return to my subject—change from one age to another. Often I have observed in my personal affairs, even those most intimate, such a flux that you would be inclined to say there was an evil genius who made it his business to turn things topsy-turvy: my money has been swept away, then increased, then diverted into other channels. Lest you believe my tongue is well-oiled or that I have suffered a lapse of memory, hear what happened to me yesterday, the effects of which still persist. And I am well aware of how many of these moods are able to be laid at the door of this changing disposition. I had dined serenely, yet from dinner on, such an odium of all the books of others and of my own, which had been published, swept over me that I could not tolerate the thought of them, nor the sight of them. Reason remains, however; I recognize the cause to be like a seizure of melancholy—and especially since these books which continue to live are the best of books!

But this affair of my moneys is another matter: it is neither trick nor madness; no one takes it away, there is no mistake; to what then is due that constant flux if not to some external confusion?

53.
QUALITY OF CONVERSATION

IN THE ART of conversation, for a number of reasons, and the more so at this time of my life, I realize that I possess little grace. First because I love solitude, and I am never more with those I love than when I am alone. For I love God and my good genius. Upon these, then, when alone, I meditate: the one who is Boundless Good, Eternal Wisdom, the Source and Creator of pure light, our true Joy without danger of loss, the Foundation of Truth, Love freely given, the Creator of all things who in himself is blessed, the Guardian and the Deep Desire of all the Saints; the Depth and Height of Justice, watching over the dead nor unmindful of the living. The other is my familiar spirit sent by his command for my defense; compassionate, my good adviser, my helper in adversity and my consolation.

What man, then, of any condition whatsoever will you point out to me who does not always go about polluted as it were, with the burden of his own filth and defiled by his own water—nay impure to his very veins? And not a few, although they may be more gracious, have, nevertheless, the belly full of worms; many of these, and many women also, whose wont it is to please—to give all things impartially their due—scratch their lice; some stink from their armpits, some from their feet and many more from their foul mouths. When I consider these facts, what mortal in the world is there that I, regarding his body, may love? Nay the little puppy, the little kid will prove much purer and cleaner.

But I turn my consideration to the souls of men; what

animal do I find more treacherous, more vile, more deceitful than man? I set aside those portions of his spirit which are subject to disorders, and would love his intellect alone. But what intelligence is sincerer than divine intelligence, or loftier, or more unassailable, or really capable of teaching truth? The libraries are stored with books, yet souls are impoverished for want of learning. Men copy; they no longer compose. Talent is not lacking, but something else. What is there, then, that I may hope for from the society of men? They are garrulous, greedy, false, and scheming. Show me, even in this century so flourishing and with opportunity so useful as the invention of printing offers, one man who has discovered even a hundredth part of those things which Theophrastus discovered, and I give him my hand. Rather, by their trifles, their *οὐ* and their *ὄν*, they confound this fair and beautiful invention.

But neither is this association with men profitable for these very ends, because inventions are the result of the tranquil life, free from annoyance and given to purposeful meditation as well as to experimentation. All of these are the product of solitude and not of the society of men, as we may read of Archimedes. Personally, of sixty discoveries I have made, perchance, I owe not even twenty to others, or to my associations. I would not wish to be branded for false should I say they were somewhat fewer. I confess that in mathematics I received a few suggestions, but very few, from brother Niccolò.[1] Yet how many have come to naught! For a large number I am indebted to other sources; yet to this very day I owe much to that secret virtue of a flash of intuition or to something even more potent.

What, therefore, have I to do with men?

Another argument is this: fortunate men disdain my society; I myself have no need for the company of the wretched, for should I desire to offer them soothing words, I could not correct their condition; if I exhibit any irritation, they take it amiss. Old men are, besides, morose, melancholy, complaining, and odious; what profit will accrue from the words of such as these?

The brevity of our mortal life is also a claim. Our days are seventy years and among the stronger, eighty. As but little is left for trifles, what part of my day should I give to these? Should I take the time of meditation? A breach of right and duty! The time of writing? Foolish, to return to concerns from which I had once been able to escape—the very murder of my leisure! Should I subtract the time from my bodily exercise, from sleep, from my domestic affairs?

Then with whom should I spend this time? With friends? Useless, for they desire the product of my labors and not conversations! With others? If with the experienced, they will perchance think they know much more than I. And if indeed they do, there will be argument concerning definition. Suppose I should wish to learn their views? or to instruct them? Impudence and extravagance will it be to impart thus freely your knowledge that you may get in return nothing but hatred!

Should I associate with one man alone or with many? If with many, to what end? If with one, think you that he will be your god? You will so move others to envy that you may be forced again to descend into the stormy sea of unpopularity. Many may address you at the same time, yet secretly they will deride you; you will expose yourself to many events with no profit to yourself.

In the count of that which is of highest importance, charm is a requisite for social intercourse, a pleasing manner for conversation—and from both of these the nature and habits of an old man are far removed. For all of which I summon no other witness than Aristotle himself. For these reasons I have ever been difficult in conversation; and this in turn became one of my excuses for not attending banquets. I do not, none the less, thrust away from myself the good, the just, and particularly those who call on my pity or who deserve well from me, nor do I deny the wise.

But, you will say, man is a social animal and what will you accomplish by renouncing the kingdom of this world in this manner? What can it do for you? You boast of having powerful friends, or is this but an empty vaunt, or wherein is it to your advantage? For there are those who give pleasure to these same friends of yours by dining with them, by gaiety and mirth; why should they desert these to take up with you? What benefits do your studies beget if others put no faith therein?

Scire tuum nihil est, nisi te scire hoc sciat alter.[2]

Finally if these relationships so useful to human nature be absent, you will undergo many discomforts; this I know can be held against me. But it does not escape me that many conditions seem hard and untenable which appear quite different when we come to deal with them, as, on the contrary, others seem easy and profitable which are otherwise difficult and unreasonable.

And since this amenity of life is generally maintained by me, that is all that is necessary, when those are taken into account whom I have said I admit to my friendship, for they are enough, and are both more useful than a host of acquaintances and also more trustworthy.[3]

54.

AND THIS IS THE EPILOGUE

WELL, THEN, I am now no longer troubled by imputation of falsehood, and am as one who has grown old in the love of truth, with which the love of God, the hope of immortal life, the possession of so much distinction and the advantage of wisdom are closely joined: which united good I am loath to confound by one false move. Let us leave that to those who are deceived in their ignorance, who rejoice in lying words, who exaggerate all they have heard, read or even seen, hoping thereby to be able to deceive. But will they, relying on some hope, make question concerning those things to the evidence of which they will subscribe not one jot of credence—though there be a thousand testimonies of one good man, or a thousand bearing witness?

Nothing, accordingly, save a love of truth influences me: but men clearly do not think alike. They differ even as birds: for there are carnivorous birds and others of foul habits, as ravens and crows, which are busy with their greed, plunder, tricks and cruelty; there are still others of somewhat better breed, as the eagle or the hawk, and these are fired too vehemently with arrogance and passion. What wonder, then, if such men give no consideration to these truths, although of such examples of fidelity to the truth the books of all historians, sacred as well as profane, are full; and instead of upholding those who are our opponents, God and his whole creation arrayed with the saints, sides with those who have true wisdom—a host as against a handful, true men against liars and the wise fools.

This one measure ought to be a concern of princes, that, in accordance with ancient precedent they should punish with just penalties those who inveigh against the just and the learned. If they have neglected this, one man from all and for all—setting aside the fact that their power may be overturned—may exact it.

For this reason, then, and not for the sake of self-glory did I in an earlier chapter array those references to my art.[1] Can it be that some think me so senseless as to wish by this means to return to the burden of my cares?[2] I did it that men, in so far as they can seek out the truth, may know me for such as I am—in a word, a teller of the truth, an upright man, and indebted for my powers to a divine spirit.

Furthermore, human affairs are accomplished through skill, reason, counsel, inspiration, or because of favorable opportunity, or on violent impulse or by mere chance: skill is that practical knowledge such as the smith possesses; counsel is subordinated to the decision of men, who if they be not friends can be moved neither by any kindnesses nor by any swift experiments in force. Inspiration is a *rara avis*, nor has it ever been to anyone a source of satisfaction in every respect. Opportunity, especially when forseen, is good but this does not always clearly appear. As reason is the safest guide on well-known ground, and tested as well, so chance is bad; violent impulse, because of which one is at times plunged into trouble, is the worst, because it is foolish. For rational concerns, my handbook *Actus* is my guide; in affairs of divine import there is that flash of intuition to which one may allude but which cannot be described; and for all things more than mortal there is my attendant spirit which can neither be described nor alluded to and is not under my control.

NOTES

PROLOGUE

1. Marcus Aurelius Antoninus, to whose famous *Meditations* or *De Rebus Suis* Cardano has reference.

2. For the careers of these men see Chapter 35 of the *Vita.*

3. Probably Flavius Josephus to whose writings Cardano so frequently refers. His life was entitled *Ἰωσηπου βίος.*

4. Galen (Claudius Galenus, 131–201), the celebrated Greek physician who in 165 A.D. settled at Rome. Girolamo Cardano liked to think he resembled Galen in many respects, and indeed he did. Both men were above the intellectual level around them; both were boasters, both believers in dreams and omens, both deficient in personal courage. Both sought to combat bad practice in medicine; both were devoted disciples of Hippocrates and wrote commentaries on the works of the Father of Medicine. Galen wrote two treatises on the books he himself had written and on the order in which they were to be taken. Those treatises he called *De Libris Propriis.* Cardano wrote three works of precisely the same kind, and gave them the same title.

CHAPTER ONE

1. Pope Urban III (1185–87) was Umberto, of the noble Milanese family of the Crivelli. He was created Cardinal by Lucius III in 1182 and Archbishop of Milan in 1185. On November 25, 1185, Lucius died at Verona, and the Archbishop was elected to succeed him on the same day. He was crowned on December first. This haste was probably due to fear of imperial interference. *The Catholic Encyclopedia.*

2. Pope Celestine IV was Godfredo Castiglione, a native of Milan, and a nephew of Urban III. He was made Cardinal by Gregory IX and suc-

ceeded him October 25, 1241, at the height of the papal warfare with the Emperor Frederick II. He died after a reign of fifteen days. *Ibid.*

CHAPTER TWO

1. Cardano in *De Consolatione*, Liber III, Opera, Tom. I, p. 619, says: "On the 24th day of September in the year 1501 I was brought forth half dead." All editions of the *Vita* give 1500. In his horoscope Cardano gives the date and hour of his birth as September 24, 1501, at 6:40, the *hora noctis prima* being six o'clock according to the old Italian method of reckoning.

2. *Ptolemæi de Astrorum Iudicium*, etc., Critical Commentary on Ptolemy's Treatise on the Heavenly Bodies, combined with Twelve Specimen Genitures, by Girolamo Cardano. This work was published in four books, together with the twelve horoscopes, the *eighth* of which is Cardano's own.

3. Mars and Saturn are the malefics.

4. Virgo is a *human* sign.

5. Cardano's horoscope shows the Sun in Libra, in which position it is said to be in its fall.

6. Augustus was born September 23, 63 B.C. In *De Utilitate* Cardano gives September 23 as his birthday, and by modern reckoning 6:40 P.M. would still be September 23, whereas under the old Italian style it would be the first hour of the following day, i.e., September 24.

7. Cardano can scarcely be referring to any but Columbus's first voyage (*classem primum emiserunt*) yet if this is the case, he has missed the date by more than seven weeks. Columbus set sail from Palos on August 3, 1492.

CHAPTER THREE

1. Chiara (Clara) Micheria: She was a young widow in her thirties with three children when she went to live with Fazio Cardano, who was fifty-six. They were not married for many years, if ever. Cardano gives no information regarding any marriage in the *Vita*, implying only that his parents, except for one period of separation, maintained a home together until the father's death.

CHAPTER FOUR

1. Cardano, about to set down an orderly and chronological account of his life, compares the method he has adopted with that of Suetonius, who does not follow the chronological order in his *Lives*, but groups facts according to their nature, Cardano's own plan, to be sure, if the *Vita* be taken as a whole. His comment, therefore, must apply to this chapter only.

2. The date of this event nearly corresponds with the date of a plague of syphilis which appeared in Naples in 1495 and swept all over Europe, invading court and camp alike. (See Fracastoro's poem, *Syphilis*). Cardano's description, however, of the effect the pestilence had on him corresponds more closely to the plague described in a work by Girolamo Mercuriale: *De Peste in Universum præsertim vero de Veneta et Patavina* (On the Plague Generally and Especially in the Localities of Venice and Padua, Basle 1576). In Chapter Twenty-Nine (*De carbunculo pestilenti*) Mercuriale mentions, among many other symptoms, "carbuncles on all parts of the body," and adds that "many of widely separated localities fell victims at the same time, German peoples at the same time as those of Italy in Venice, Padua, *Milan*, and Calabria." He does not, however, mention any plague in Milan in 1501, but says: "There was a great epidemic in 1528 when, according to the most noted writers, a third part of the people perished." This plague Cardano mentions later in this same chapter.

3. Old Fazio Cardano was wont to declare that he, like Socrates and others, was favored with the society and advice of a familiar spirit or demon which attended him for thirty-eight years.

4. On May 14, 1509, the French defeated the Venetians near the River Adda, and celebrated their triumph in Milan.

5. Accademia Ticinensis, now the University of Pavia. It was founded in 1361 on the site of an old and famous law school which had existed there from the 10th Century. The present building dates from 1490, thirty years before Cardano entered, with considerable later additions. Columbus is said to have been a student at Pavia at one time.

6. Padua was for centuries a center of culture, and its university, founded in 1222, rivaled that of Bologna in attracting students from all parts of Europe. Galileo was a teacher there in 1592–1610.

7. The Venetian Degree in Arts: This degree was not in high favor in Italy; it was sought by those who could not afford much expense for schooling.

8. On many occasions and in many works Cardano affirms his rectorship, and just as often and in as many places he asserts his devotion to the truth, so the statement that he was rector must be accepted, although in the extant records of the University of Padua there is no mention of his name in the lists of rectors. From 1508 to 1515 the University had been closed because of the wars. From 1515 to 1525 no rectorships are recorded, and for 1526 only a pro-rector was appointed. Cardano's term, then, must have been an irregular or pseudo rectorship not formally acknowledged.

9. College of Physicians of Milan: i.e., Medical Association.

10. Xenodochium: the poorhouse of the workhouse. An establishment whose doors were open to the houseless stranger, maintained from religious motives by various communities in obedience to admonition in Matthew 25:34–40.

11. Prelate Archinto: Filippo Archinto, Cardano's first patron, a young man of excellent birth and breeding, of wealth and wisdom. He was Cardano's life-long friend. Later he became Bishop of Milan.

12. Brissac, Marechal Cosse, high officer to the King of France. Some years later he offered Cardano a thousand crowns a year to become, ostensibly, his physician; but what he really sought to have at hand was the physician's mathematical ingenuity. Brissac, albeit a gallant, had taste and scholarship, and quickly recognized Cardano's talents. Lodovico (Louis) Birague was commander-in-chief of the forces of the French king in Italy.

13. Francesco della Croce: For his part in Cardano's career see Chapter 40 of the *Vita*.

14. Cardinal Morone: Cardano's life-long friend and patron. He was a man of great learning, became Cardinal, stood in high favor with the Pope, and was in 1546 president of the Council of Trent, with which his name is famously connected.

15. Pope Paul III, Alexander Farnese. As Cardinal he had begun to build the splendid Farnese Palace at Rome.

16. Prince d'Iston: Alphonso d'Avalos. He was a liberal patron of arts and letters, a noted gallant and a great dandy, and very popular as Governor of Milan. Cardano mentions him again as host to a hunting party at which the physician was a guest.

17. Vesalius: Andreas Vesalius, the founder of modern anatomy, and a pioneer of comparative anatomy and race craniology. He was born at Brussels in 1514, studied the classics at Louvain, and medicine at Cologne, Paris, Louvain, and Padua whither he went attracted by the fresher and more liberal scientific spirit of North Italy. He lectured on anatomy at Louvain, Venice, and Padua, where he was Professor in 1537–44, and at Pisa and Bologna. In 1544 he was made physician-in-chief to Charles V and after the latter's death, to Philip II at Madrid. After a pilgrimage to Jerusalem (undertaken at the Emperor's suggestion to expunge some experimental indiscretions in anatomy!) he was invited to Padua to occupy again the chair of anatomy, but died on the way home of exhaustion following a shipwreck, in 1564.

18. To cure the asthma of John Hamilton, Archbishop of St. Andrews and brother to James Hamilton, Regent of Scotland.

19. Henry II of France and Emperor Charles V were then at war. Cardano was a subject of the emperor.

20. Cardano probably means Mary of Guise.

21. Cardano's son Giambattista was executed for murder.

22. To become a member of the faculty of medicine at the University of Bologna, reputed to be the oldest in Europe.

23. Cardano does not tell us exactly why he was imprisoned, but it was probably on a charge of impiety. He had given bond to the extent of 1,800 gold crowns as security for being his own jailer.

24. Don John of Austria annihilated the Turkish fleet at Lepanto on October 7, 1571.

25. Gregory XIII.

CHAPTER FIVE

1. There are two well-known portraits of Cardano. The one found facing the title page of Spon's great 1663 edition of Opera Omnia shows

a man of pleasing and serene features, and answers rather well the description Cardano has left us of himself. It was made in his 68th year. The other, a wood cut, which appears on the title page of Morley's *Life of Girolamo Cardano*, shows a much younger man with a somewhat troubled countenance. Cardano was thirty-eight when this was made, and bitter because of the disesteem in which his scholarship was held. Around the medallion runs the legend: "In Patria Nemo Propheta Acceptus." It was originally prefixed to his book, *Practica Arithmeticæ.*

CHAPTER SIX

1. Cardano omits the long illness in his eighth year which was caused by eating green grapes. See Chapter 4.

2. Asclepiades, of Bithynia, was a physician at Rome in the 2d Century A.D. One of his treatments was to place the patient in a sort of huge cradle rocked up and down by slaves.

3. A repetition of statements made earlier in this same chapter with slight variations of detail.

4. Pleasure consists in the absence of pain, i. e., the *indolentia* of Cicero.

5. *Laboravi etiam amore heroico:* perhaps *Weltschmerz!*

CHAPTER EIGHT

1. *Squalus:* a kind of ray fish not otherwise identified.

2. Perhaps Cardano is referring to an Italian game of his own day, but remarking his devotion to everything Galen had to say, one is reminded of Galen's *Περὶ τοῦ διὰ Μικρᾶς Σφαίρας Γυμνασιου* (On Ball-Playing as an Exercise), a masterly statement of the true principles of exercise. The best of all exercises, he says in this treatise, are those which combine bodily exertion with mental reaction, such as hunting and ball-play. But ball-play has this advantage over hunting that its cheapness puts it within reach of the very poorest, while even the busiest man can find time for it. It exercises every part of the body, legs, hands, and eyesight alike, and at the same time gives pleasure to the mind.... There is no doubt that Cardano had this treatise in mind.

3. It is not clear just what Cardano means by this "exercise" which he calls *politura chartarum*, and which Hefele translates "*Das Polieren von Platten.*"

4. Cardano becomes so rashly enthusiastic about his system of *fifteens* that he neglects to take account!

5. Henry Morley in his biography of Cardano says, "Jerome amused himself with the manufacture of a little burlesque sketch of the philosophy of victuals, which may be taken as a satire upon some of his own graver generalizations." The old physician was no doubt "writing after supper, with a twinkle in his eye."

CHAPTER NINE

1. Herostratus, the Ephesian, who, in order to immortalize himself, set fire to the temple of Diana at Ephesus on the same night that Alexander the Great was born, 356 B.C. The Ephesians passed a decree condemning his name to oblivion, but this only increased his notoriety and certainty of obtaining his object.

2. Horace, Odes III, 29, line 1 and lines 41–48.

Tyrrhena regum progenies: (Mæcenas) "sprung from Tuscan kings" is the first line of the ode. Dryden's translation of lines 41–48 is as follows:

> Happy the man, and happy he alone,
> He who can call to-day his own:
> He who, secure within, can say,
> To-morrow do thy worst, for I have lived to-day!
> Be fair or foul, or rain or shine:
> The joys I have possessed, in spite of fate, are mine;
> Not Heaven itself upon the past has power,
> But what has been, has been, and I have had my hour.

3. "To be content that times to come should only know there was such a man, not caring whether they knew more of him, was a frigid ambition in Cardano." Sir Thomas Browne: Urn Burial, 5.

CHAPTER TEN

1. Catullus, 8, 3:

> So bright, so white, the suns that shone before . . .

2. Boasts Cardano's house of this great Cardano's name,
Whose wisdom wide surpasseth all his age for fame.

CHAPTER TWELVE

1. Andrea Camuzio, one of the most illustrious physicians of the period.

2. Duke of Milan: Francesco Sforza II, the last of the Sforzas.

3. Mary of Hungary, sister of the Emperor Charles V, Queen of the Netherlands and Ruler of the Low Countries. She was called to Belgium by her brother, where she reigned for over twenty-four years at Brussels with great intelligence and firmness.

4. Garamantes: Herodotus, in Book IV, 183, where he describes the nation of the Garamantes, does not make this statement about *them*, but rather, in ¶ 184, about the *Atarantes* who "dwell ten days' journey from the Garamantes." His remark is, "These people [i. e., the Atarantes], when the sun is exceedingly hot, curse and most foully revile him, for that his burning heat afflicts their people and their land."

CHAPTER THIRTEEN

1. *Γνῶθι σε αὐτόν*: "Know thyself," the saying of Thales of Miletus, one of the seven wise men.

2. "Yet vengeance is sweeter to me than life itself."

3. "Our nature is prone to evil."

4. Like many of the humanists and their disciples, Cardano knew Greek literature largely through Latin translations. It is doubtful whether he ever read Aristotle in the original, thoroughgoing student of Aristotle though he was. His citation of Aristotle in the Latin rather than the Greek seems to point to this fact: "Homo solitarius aut bestia aut deus" (The solitary man partakes of the nature of the beasts or of the gods). In Chapter 53 he discourses on the reasons why he prefers solitude.

5. Ranconet more than any other Frenchman won Cardano's deepest affection, and he declared that this friendship alone made worth while his visit to France.

CHAPTER FOURTEEN

1. Marco Antonio Bragadino: a Venetian General (1520–1571). He was governor of Cyprus when the Turks besieged it. He made an extraordinary defense and when finally food gave out and only seven casks of powder were left, Bragadino surrendered, but obtained from Mustafa Pasha an agreement whereby the people of Cyprus were to retain their liberty. The garrison and leaders were to be sent to Venice, but Mustafa, breaking his word, took Bragadino to Constantinople, had him flayed alive, and sent his skin to various cities in Asia Minor. This inhuman treatment is attributed to Bragadino's unwillingness to abjure the Catholic faith.

2. The grief to which Cardano is evidently referring here and in the following incoherent lament is the profound despair which he suffered upon the execution of his eldest son, Giambattista. Mad with the pain of it, he was never able thereafter to refrain from including in almost everything he wrote wild expressions of his sorrow.

The miracle of the emerald is recounted in Chapter 43 of the *Vita.*

3. Cardano burned many books and monographs, left 138 printed, and 92 in manuscript, according to the *Vita.*

4. In 1562 the Accademia degli Affidati was founded at Pavia, and at once became famous throughout all Italy. Within a few years its membership included the Cardinals of highest rank, some of the princes and kings of Europe, and the Emperor himself, Philip II.

5. Horace: Epistulæ I, 1, 41: "To fly from vice is virtue."

6. Matteo Corti: Professor of the Theory of Medicine at Padua, 1524–1530. He was Cardano's teacher at one time, and encouraged him constantly. He lectured also at Florence, Bologna, and Pisa.

7. It was to this statement that Gabriel Naudé, Cardano's arch-critic, and editor of the *Vita*, took such exception. *De Vita Propria*, however, seems to bear out the accuracy of Cardano's declaration. Except for minor discrepancies, plainly accidental, he is utterly consistent.

CHAPTER FIFTEEN

1. A college friend of Cardano's with whom he had been wont to discuss the possibility of a future existence. Years after Cardano declared

that he enjoyed a spiritual visitation from this friend who had died while still a young man.

2. Another college friend. While still a student Cardano had written some fifty sheets of mathematical commentaries, which this Ottaviano lost for him. Later Scotto, inheriting a printshop in Venice, became his first publisher.

3. See Chapter 40 (1) of the *Vita* for an account of the cure of this man.

4. Francesco Sfondrato: One of Cardano's firmest and most powerful friends. As senator he was influential with the Emperor and later became Governor of Sienna and Cardinal. For Cardano's account of the cure of his son see Chapter 40 (1) of the *Vita.* Cardano forecast Sfondrato's horoscope in "De Exemplis Centum Geniturorum LXXX." See Opera Omnia. Tom. V, p. 495.

5. Andrea Alciati: The great jurist of his age (1492–1550). Introduced the study of Jurisprudence into France by a series of lectures at Bourges, 1528–32. He was one of the Professors at the University of Pavia when Cardano was summoned to the Chair of Medicine there.

6. Francesco Alciati: Heir of the preceding. He secured Cardano's appointment to the University of Bologna in 1562, and was faithful to the end of Cardano's life, to the close friendship existing between them.

7. Giovanni Morone: One of the most notable men of Milan. His father had been chancellor to the last Sforzas, and later had diplomatically adjusted his relations to Bourbon. His son was Cardinal Morone. See Chapter 4, note 14.

8. Pietro Donato Cesio: A patron of Cardano's later life. He was his protector and supporter at Rome.

9. Fernando Gonzaga: Successor to D'Avalos. He became Governor of Milan in 1546, and, though a harsh man, was always friendly to Cardano.

10. Cardinal Borromeo: One of the finest and most upright characters in the church and state in the 16th Century. Cardano had cured his mother, and Borromeo never relaxed his kindly protection over the physician. He was later canonized for his great virtue. See also "Eloges des Eveques," Antoine Godeau, Paris, 1665. Eloge 98.

"It may be said of St. Carlo Borromeo that he was an abridgement of all the bishops given by the Lord to his church in the preceding ages; and that in him were collected all the episcopal virtues that had been distributed among them."

11. Sir John Cheke (1514–1557) "taught Cambridge and King Edward Greek." Regius Professor of Greek at Cambridge in 1540. He was then Fellow of St. John's, and afterwards Public Orator, and Provost of King's. In 1542 Roger Ascham, writing on the flourishing state of classical studies in Cambridge, said, "it is Cheke's labours and example that have lighted up and continue to sustain this learned ardour." See Sandays: *A History of Classical Scholarship.*

12. It was at Buonafede's suggestion that Cardano established his first practice at Sacco.

13. Guillaume Duchoul: French antiquarian of Lyons whose works on military engineering and discipline among the Romans, and on Roman religion were translated into Italian, Latin, and Spanish.

14. Giampietro Albuzzio: Fellow-professor with Cardano at Pavia. He had also, like Cardano, started his career at Gallarate; at the age of twenty-five he was invited to lecture on Rhetoric at Pavia, where he stayed for forty years, though he received many calls to other universities. He was a close friend in sorrow, for it was to Giampietro Albuzzio that Cardano poured out his grief after his son's execution. To him also he dedicated a book, "De Morte," a dialog upon a sorrowful theme.

15. Matelica: one of the small states of the church.

CHAPTER SIXTEEN

1. "He had a rugged love of truth and justice; he remembered benefits, and when affronted could afford deliberately to abstain from seizing any offered opportunity of vengeance. He governed his pen better than his tongue, and carefully restrained himself from carrying into his books the heat he could not check in oral disputation. He left his enemies unnamed, and though he now and then is found devoting some impatient sentences to writers who had treated his opinions rudely, yet it seems at first sight absolutely wonderful that a man so sensitive and so irascible, so beset by harsh antagonists as the weak-bodied Jerome, should have filled so many volumes with philosophy

and so few pages with resentment. *The wonder ceases when a closer scrutiny displays the difference in intellectual and moral weight between Cardano and most of his opponents."*

Henry Morley, *Girolamo Cardano*, Vol. I, p. 128.

CHAPTER SEVENTEEN

1. University of Bologna: Said to have been originally founded in the 5th Century. It acquired a European reputation as a School of Jurisprudence under Irnerius who introduced the study of Roman Law in the 11th Century, and attracted students from all Italy and western Europe, the enrollment being from three to five thousand. The study of medicine and philosophy was later introduced; a faculty of Theology was established by Pope Innocent VI. Anatomy of the human frame was first taught here in the 14th Century. Galvanism was discovered here by Galvani in 1789. It is a remarkable fact that the University of Bologna has had several women professors. Laura Bassi lectured on mathematics and physics; Madame Manzolina on Anatomy, and from 1794 to 1817 Clothilda Fambroni was professor of Greek. The present buildings, begun in 1562 when Cardano was invited to Bologna, were erected chiefly through the generosity of Cardinal Borromeo. They are still standing.

2. "Certainly," says Henry Morley in his biography of Cardano, "there was no lack of rivalry and heartburn among the professors who were in too many cases emulous and envious of each other. Cardano had a great name and not a winning nature." Vol. II, p. 280.

CHAPTER EIGHTEEN

1. Cardano's reference to these examples of the jeweler's art reminds one that though Cardano and Benvenuto Cellini were contemporaries their paths apparently never crossed. One would like to think Cardano's jewels, and wrought gold and silver articles were the handiwork of the Florentine.

2. Luigi Pulci, Florentine poet: 1432–87, Cardano's favorite.

CHAPTER NINETEEN

1. In Cardano's *Liber De Ludo Aleæ*, a treatise on games of chance, he declares that the stake should be the only excuse for gambling;

otherwise there is no compensation for loss of time, thought, etc., which could be far better engaged in other diversions.

CHAPTER TWENTY

1. Horace, Satires: I, 3, 9–15.

> One mass of inconsistence, oft he'd fly
> As if the foe were following in full cry,
> While oft he'd stalk with a majestic gait,
> Like Juno's priest in ceremonial state.
> Now he would keep two hundred serving men,
> And now a bare establishment of ten.
> Of kings and tetrarchs with an equal's air
> He'd talk; next day he'd breathe the hermit's prayer:
> "A table with three legs, a shell to hold
> My salt, and clothes, though coarse, to keep out cold."
>
> *Conington.*

CHAPTER TWENTY-TWO

1. In 1494 Charles VIII of France undertook a campaign against Naples, and thereby plunged the Italian states into a long series of wars. The Milanese territory was harassed by tumults and battles. In 1525 the battle of Pavia made the Emperor Charles V arbiter of the fortunes of Italy. These political vicissitudes, together with the vast upheaval of the Reformation, made up the troubled background of Cardano's century.

2. Avicenna: Abu Ali al Hosain Ibn Abdallah, called also Ibn Sina, and in Chapter 45 of the *Vita* Cardano calls him Hazen. Avicenna, the Arabian philosopher of the 11th Century (980–1037) holds the foremost place in Arabian medicine. His "Book of the Canon of Medicine," when translated into Latin, overshadowed the authority of Galen himself for four centuries. He taught at Ispahan, combining instruction in medicine with the exposition of Aristotle.

CHAPTER TWENTY-THREE

1. Paul of Ægina, a celebrated Greek physician and medical writer of the seventh century A.D. Few details of his life are known except

that he traveled much, thereby acquiring the reputation of "medicus ambulans." Of his works, one is extant: De Re Medica Libri VII.

2. This story is supposed to illustrate his adage, "multa modica faciunt unum satis." The crumbling bits of plaster brought on the collapse of the whole ceiling!

3. The Latin texts are all very obscure and very much confused at this point, and it is quite possible that the interpretation of the entire paragraph is therefore at fault.

CHAPTER TWENTY-FOUR

1. This street still bears the same name as in Cardano's day, and may be found near the Porta Ticinese.

2. The Via dei Maini no longer exists. It was possibly in that region around the castle (seat of the Visconti and Sforza) which gave place to the park or the arena built by Napoleon I.

3. Formerly in the southeast quarter of the town of Pavia not far from the Arsenal.

4. S. Maria di Vénere: Possibly located on the site of an early temple of Venus, whose cult was popular in Pavia before the Christian Era.

5. S. Maria in Pertica: in northeast quarter near the castle at Pavia.

6. This may be the same as the present Via Gombruti, running north and south just west of the center of the city of Bologna.

7. In southeast quarter of Bologna.

CHAPTER TWENTY-SIX

1. A full account of Aldobello Bandarini, Cardano's father-in-law, is found in Book iii of *De Utilitate ex Adversis Capienda.* He was an innkeeper, a man-about-town, something of a swashbuckler and military roustabout, and given to lavish, and perhaps noisy hospitality. This may account, perhaps, for Cardano's misgivings when he saw mine host Bandarini and his numerous progeny moving in next door.

2. This thought need not reflect too darkly upon Girolamo's character in view of the times. Compare the *Autobiography of Benven-*

uto Cellini (1500–1571) for the license of sixteenth-century life in Northern Italy.

3. "Mirum dictu ut flatim e gallinaceo factus sim gallus." Cardano gives a full account of his experience in *De Utilitate ex Adversis Capienda*, Book II, Chapter X: De Veneris Impotentia. Tom. II, p. 76 of the *Opera Omnia.* The citation given is from *De Libris Propriis.*

4. Cardano's views on marriage may be found in *De Uxore Ducenda*, Chapter III of Book IV of *De Utilitate ex Adversis Capienda.* Tom. II, p. 237, Spon's Edition.

CHAPTER TWENTY-SEVEN

1. Crantor: a philosopher of Soli, among the pupils of Xenocrates, 300 B.C. From one of his works, *Περί Πένθους*, Cicero drew largely in writing the lost treatise, De Consolatione, on the death of his daughter Tullia.

2. Philandrier: French architect, scholar, and art critic, 1505–1565. The work referred to here is his *Annotationes in Vitruvium.*

CHAPTER TWENTY-NINE

1. John Hamilton, natural brother to Earl of Arran, James Hamilton, who was made regent of Scotland in 1543 to act during Mary's minority. In 1547, this John Hamilton was appointed Archbishop of St. Andrew's, and virtually manager of all that was difficult in Scottish affairs in that period.

2. For Birague and Brissac see Chapter 4. In Chapter 32 Cardano says Brissac wanted to employ him, not as a physician but as a mathematician and engineer.

3. Orontius: M. Fine, a noted French mathematician of the 16th Century; known also as *Oronce*, although the Latinized form of his name is more common.

4. Legrand was one of the physicians of the French king; Cardano, Latinizing his name, calls him *Magienus.*

5. Fernel was professor of medicine in Paris and first court physician. Jacques de la Boë was Vesalius' teacher. Cardano calls them Pharnelius and Silvius.

6. The title here referred to is *Defender of the Faith*, which Cardano refused to acknowledge as a prerogative of Edward VI, inasmuch as he himself was a faithful subject of the Pope.

CHAPTER THIRTY

1. Perhaps this made an impression on Cardano as the one humiliating or embarrassing experience in his otherwise triumphal return journey from Scotland to Milan, on which he had received highest honors at every turn.

2. The Wolf: Possibly Delfino, "hic nothus."

The Fox: Cardano's rival or concursor, the "vir callidus."

The Sheep: Fioravanti, the tool of the other two.

There is some confusion in Cardano's application of these epithets, so that it is not strange Morley's account does not seem accurate.

3. In a more elaborate account of this incident in his Paralipomena, Liber III, Chap. VI, Cardano adds that the ring was squared so that it could not roll. (Opera Omnia, Tom. X, p. 459.)

4. Rome was demolished by the Constable of Bourbon, May 6, 1527. Within a year and a half, the Eternal City had lost two-thirds of its population. It was the remaining years of the century rebuilding, and through the successive efforts of Popes Paul III, Julius III, Paul IV and Pius IV rose again a new city, yet even while Cardano lived there, from 1570 to 1576, it must have presented many devious ways to the unwary.

CHAPTER THIRTY-ONE

1. Marcus Æmilius Scaurus—became stepson of Sulla, known for the lavish entertainments he gave while Curule Aedile in Rome in 58 B.C.

Seneca, the Philosopher, tutor and adviser to Nero.

Manius Acilius Glabrio: judge who presided at impeachment of Verres in 70 B.C. Consul in 67, and subsequently the successor of L. Lucullus in command of the war against Mithridates, in which he was superseded by Pompey.

2. *Diepam iuxta Drudim fluvium*:

If Cardano means Dieppe—though just why Dieppe is hard to

say—there is something wrong with the geography, for the Arques near Dieppe was never, apparently, called the Drudis.

CHAPTER THIRTY-TWO

1. Referring to the illustrious records of various members of the Brutus family. Cf. the legend of Lucius Junius Brutus.

Gaius Mucius Scævola who burned his right hand in the flame in the presence of King Porsena by whom he was then freed for his firmness.

Gaius Fabricius, surnamed Luscinus: sent as a legate by Roman Senate to King Pyrrhus in 281 B.C. to treat for terms. Pyrrhus is said to have endeavored to bribe him by large offers which Fabricius, poor as he was, rejected with scorn.

2. Henry II, then at war with the Emperor Charles V.

3. From the English court which Cardano was visiting (under no contract) at that time. The royal hand was that of Edward VI.

4. Morley questions whether Cardano may not mean "Dobbin" when he says, "asturconem obtulit, Angli eum appellant patria lingua *Obinum.*"

5. Then Governor of Milan.

6. Cardano was a subject of the Emperor then at war with France.

7. What honors these may have been—at Padua and at Venice—I have been unable to discover.

8. Chapter 47 gives a full account of this "*Spiritus*" which Cardano claims attended him.

9. In Chapter 48 Cardano says that Andrea Alciati was wont to call him *Vir Inventionum.* That is, possibly, the *cognomen* he is thinking of.

CHAPTER THIRTY-THREE

1. You invite new indignities by tolerating the old.

2. Horace: Sermones I, 6, 81. (My father) himself, a guardian true and faithful, was at my side even in the presence of my teachers. . . .

3. *Diarob cum Turbith*: Turpeth.

4. For unicorn's horn see *Bestiary* of Philip de Thaun, in which the unicorn received his due praises. In *Vulgar Errors*, Book III, 23, Sir Thomas Browne discusses the questionable virtues of the horn.

5. Horace, Sermones II, 8, 95:

> *Velut illis Canidia adflasset peior serpentibus Afris.*
>
> As if Canidia's mouth had breathed an air
> Of viperous poison on the whole affair.

Cardano has substituted *me* for *illis* in his version of the citation.

6. Pliny, Nat. Hist. XI: "oculos . . . serpentium catulis et hirundinum pullis, si quis eruat, renasci tradunt." (It is said that if any one destroys the eyes of the young of serpents and of swallows, they grow again.) "in ventre hirundinum pullis lapilli candido aut rubenti colore, qui chelidonii vocantur . . . reperiuntur." (In the crop of the young of swallows small stones are found of a white or red color which are called swallow-gems.)

7. Perhaps Cardano is thinking of the following Aphorism: "Fever supervening on painful affections of the liver removes the pain."

CHAPTER THIRTY-FOUR

1. Cardano, though he had some knowledge of Latin in his boyhood, never learned to write in that language until he was nineteen or twenty years old.

CHAPTER THIRTY-FIVE

1. Lodovico Ferrari: For the romantic story of this young man's career with its truly Italian flavor, see Morley's *Life of Cardano*, Vol. II, 265–269. Cardano himself wrote a sketch of the young man's life. It was published for the first time in his Opera Omnia, 1663. It may be found in Tom. IX along with the Life of Alciati.

2. Brief is life, old age is rare
   To him who flings restraint away;

Then pray that all you prize as dear
May shun satiety's dull day.

CHAPTER THIRTY-SEVEN

1. Juvenal: Sat. I, 15.
I then have withdrawn my hand from wielding the ferule.

CHAPTER THIRTY-EIGHT

1. To keep him from coming to Bologna; see Chapter 30.

CHAPTER THIRTY-NINE

1. The prejudices of the age against the newer methods of anatomy were violent, as the career of Andreas Vesalius testifies. His treatise *The Fabric of the Human Body* was attacked by all the conservatives in the field of medicine, and his practical experiments shocked society.

2. Intellectual interest directed toward many points is less efficient for concentration on one or another single point.

3. Martial: 4, 29, 8.
Persius' one book's more celebrated far
Than Marsus' bulky Amazonian War.

4. Horace: Odes III, 30. The word order demanded by the meter is . . . *Scandet cum tacita virgine pontifex.* Herbert Grant's translation is as follows:
Long as the priest the Capitol ascends
And her chaste steps the silent vestal bends.

CHAPTER FORTY

1. *Πυρετὸν ἐπὶ σπασμῷ βέλτιον γένεσθαι ἢ σπασμὸν ἐπὶ πυρετῷ. Convulsioni febrem advenire præstat quam convulsionem febri.* It is better that a fever succeed to a convulsion than a convulsion to a fever. Hippocrates: Opera V, III, p. 715.

2. Duke of Sessa, Gonsalvo Ferrante di Cordova, Governor of Milan, appointed by Philip II to succeed Ferrante Gonzaga in 1558.

3. Aëtius, of Amida in Mesopotamia, a Greek physician of the 6th Century A.D. who lived in Constantinople as imperial physician. He was the author of a great miscellany on pathology and diagnosis in sixteen books.

4. Oribasius, a native of Pergamum, and, in the 4th Century A.D., physician and adviser of the Emperor Julian the Apostate. His medical works comprised seventy-two books of which twenty-two are preserved. He was banished by Valens in 363.

5. Pliny, Natural History: 7, 25:25, ¶ 92.

CHAPTER FORTY-ONE

1. *Charcas*: an early name for Bolivia.

2. *Parana*: the region in central South America west of the Parana River, approximating the location of present day Matto Grosso. Cf. location of present state of Paraná, Brazil.

3. *Acutia*: name given to a region in central South America east of Parana River. Corresponds to present Goyaz or Minas Geraes.

4. *Caribana*: on the Equator between Amazon and Orinoco Rivers, a little south of modern Guianas. Named from Indian tribes living there.

5. *Picora*: a region south of the Amazon in the very heart of South America.

6. *Quiniram partem occidentaliorem*: Can he mean *Quivira*? Bohun's Geographical Dictionary of 1691 says: "Quivira is a province in North America between New Mexico, Mount Sual(?) and Florida, which was never conquered by any of the European nations, nor indeed thoroughly discovered. In Mercator's Atlas (1633) Quivira Regnum is found on the west coast of North America about in the locality of present British Columbia, which might account for the expression "partem occidentaliorem."

7. *Cortereal*: southern Labrador, named from Caspar Corterealis, a Portuguese gentleman supposed to have discovered this region about 1500.

8. *Estotilant*: northern Labrador, which "seems to have derived its name from its lying more easterly than the rest of the province." *America* by John Ogilby, 1671.

9. *Marata*: a region, according to the Atlas of Mercator, in northern Mexico lying between Nova Granada and Nova Hispania.

10. *Antiscians*: *ἀντίσκιοί*: people on the other side of the equator, whose shadows are cast in the opposite direction.

11. *Laponiam*: thus in all editions, yet I have taken the liberty of making it Iaponia, i.e., Japan, discovered in 1542.

12. *Binarchia*: not known, unless the "double kingdom" may refer to China.

13. An echo of the famous Ode (I:3) of Horace:

> nil mortalibus ardui est
> cælum ipsum petimus stultitia, neque
> per nostrum patimur scelus
> iracunda Jovem ponere fulmina.
>
> Nay, scarce the gods, or heavenly climes
> Are safe from our audacious crimes:
> We reach at Jove's imperial crown,
> And pull the unwilling thunder down.
>
> *J. Dryden.*

CHAPTER FORTY-TWO

1. Returning from Scotland, through London in the summer of 1552, Cardano waited on King Edward VI, whose horoscope he was professionally obliged to set down. The predictions therein failed for the most part, a fact which Cardano excused on the grounds that the state of affairs at the English Court, altogether involved in the insidious intrigues of Northumberland, had warned him that a mere perfunctory statement would be the only diplomatic move.

2. For the Lament, see Chapter 50.

3. Cyprus was lost to the Turks in 1570, an event which hastened the famous battle of Lepanto, 1571.

4. Perhaps he refers to Tunis which in 1574 Don John of Austria lost to the Turks.

5. If I am reading the same treatise of Aristotle on the power of prevision to which Cardano refers, I find somewhat contradictory

statements: "This power of prevision, then, occurs in any ordinary person, and not in the wisest." Aristotle's Psychology: On Divination: Chapter II.

6. doctrina crassa
dilemma
τρόπος
amplificatio
splendor
singularis
dialectica

To this group of rhetorical "principles" or devices Cardano constantly refers. Of the first I have found no definition. *Dilemma* is the double proposition, or argument in which an adversary is pinned between two difficulties. The third is, of course, the same, *trope. Amplificatio* is Aristotle's *αὐξητικά.* Its object is to increase the rhetorical effect and importance of a statement by intensifying the circumstances of an object or action. *Dialectica* is logical discussion by way of question or answer as in Aristotle. *Splendor* is a term of Cardano's, possibly for intuition or direct insight.

7. Alexander Sforza di Santafiore, uncle of Pope Paul III. Cardinal from 1565.

CHAPTER FORTY-THREE

1. Romans 4:18 and 22.

2. *Liber Quintus Theonoston seu Hyperborarorum*; the book is a dialogue, and its theme is the life and happiness of souls after death.

3. Pope Pius V.

4. Not only is the Latin somewhat ambiguous at this point, but the example of a miraculous visitation is introduced in the form of a mere note, as if Cardano expected to expand further both the incident and the theme.

CHAPTER FORTY-FOUR

1. Cardano's mathematical treatises are the only works out of the great bulk of his published writings that have enjoyed any lasting attention. "Cardano's Rule," the general solution of the cubic equation, is mentioned—if not expounded—in modern algebras, and

discussed by learned mathematical societies (e.g., Memoires de la Societé Royale des Sciences de Liège, 1900, contains a discussion of "le cas irreductible de la formule de Cardan"). This "rule" was really the discovery of Niccolò Fontana, called Tartaglia, but it was through Cardano's efforts that it became a part of the body of mathematical science. Morley's biography of Cardano, Chapters 12 and 13, tells a fascinating story of Cardano's mathematical piracy. Professor Florian Cajori of the University of California gives us, however, a far less sentimental statement of the case in his *History of Mathematics*, an account which leaves Cardano rather in umbrage for his conduct in the transaction. I fall rather between the two. "It was the practice in those days, and for two centuries afterward," writes Mr. Cajori, "to keep discoveries secret, in order to secure by that means an advantage over rivals by proposing problems beyond their reach." From this practice I find Cardano generally free, and he seems to have acted purely (if piratically) in the interests of science, whereas it was not to be expected that a man of Tartaglia's rather crabbed nature would act thus disinterestedly.

2. A most curious statement, if I have rightly interpreted "velut cur mille aleæ in mille iactibus, si sint legitimæ, necessario iaciunt monadem."

3. *Liber de Aqua et Æthere.*

4. There exists a work supposedly of Galen entitled De Urinis, but it is of doubtful genuineness; also two other treatises, *De Urinis Compendium*, and *De Urinis ex Hippocrate, Galeno et aliis quibusdam*, which are spurious.

5. Andrea Alciati; see Chapter 48 for this fact.

CHAPTER FORTY-FIVE

1. The Defense, entitled *Actio Prima in Calumniatorem*, was Cardano's reply to Julius Cæsar Scaliger's furious attack (in his *Exercitations*) upon *De Subtilitate*. Cardano never mentioned Scaliger's name in his reply, the dignified and learned character of which gave him by far the best of the argument.

2. *Hyperchen*: The significance (overflow?) of this title is doubtful; the treatise is on an abstract philosophical theme.

3. Cardano, in this essay on Socrates, appears to have taken the role of detractor, as if for the sake of an exercise in syllogizing, just as he espoused the cause of Nero in his *Encomium Neronis.*

4. *Oratio ad Cardinalem Alciatum, sive Tricipitis Geryonis, aut Canis Cerberi.*

5. The greater part of this commentary was on the medical works of Hippocrates. In Cardano's opinion they were his best work, on which, together with his Arithmetic, he rested his assurance of lasting fame. He says he wrote them "to increase health among men."

6. Selected cases from those contained in Hippocrates' *Three Books on Epidemics.* The latter is now available in English in the Loeb translation, by W. H. S. Jones.

7. Such a book does not appear in the Opera Omnia. There is, however, a dialogue entitled *Antigorgias Dialogus seu De Recta Vivendi Ratione.* It is a single writing, not in five parts as the list indicates, but there are five interlocutors: Socrates, Chærephon, Gorgias, Polus and Callicles. Opera Omnia Tom. I, p. 641.

8. Many of these manuscripts were printed for the first time in Spon's edition of the Opera Omnia (Lyons 1663), and many of them are apparently lost.

9. *Memorialis* is the title he gives the book, but it has apparently not been preserved, at least under that title.

10. *Ars Medica: Τέχνη Ἰατρική*: of all of Galen's works none was so much studied and commented upon as this; it was a general outline of Medicine.

11. Hasen: Avicenna.

12. *De Malo Recentiorum Medicorum Medendi Uso:* Cardano's first published book. He said of it later: "I am ashamed to acknowledge that there are more than three hundred errors of my own making in the book, not counting the misprints."

13. *Hieronimi C. Cardani Practica Arithmeticæ et Mensurandi Singularis.* Printed by Bernardo Caluscho, Milan, 1539. A woodcut of the author was prefixed with the legend "In Patria Nemo Propheta Acceptus," and the work was prefaced by a verse about the book in

alternate hexameters and pentameters as well as an advertising notice about himself to far-off scholars. This attracted the attention of Petreius of Nuremberg who subsequently printed Cardano's Judicial Astronomy and his Algebra (*Ars Magna*), and reprinted his *De Consolatione*, through the courtesy of one Andrew Osiander who had taken an interest in Cardano's work.

14. Imprisonment in Bologna, perhaps for something he had written.

15. Iliad VI, 236: "gold for bronze, the worth of an hundred oxen for the worth of nine."

16. *Liber Artis Magnæ sive de Regulis Algebraicis*, 1545.

17. The two books eventually grew to fifteen: *Quindecim Libri Novæ Geometriæ*.

18. *Musica*: Divided into 5 Books: (1) General rules and principles; (2) Ancient music, rhythms, hymns, choruses and dances; (3) Music of Cardano's own time; (4) Mode of composing songs and counterpoint; (5) Structure and use of musical instruments. The style is said to be individual and interesting.

19. *De Animi Immortalitate*: "a collection of extracts from Greek writers which Julius Cæsar Scaliger with justice calls a confused farrago of other men's learning." Waters says that it is the one work of Cardano's in which materials for a charge of impiety might most easily be found.

20. *Proxeneta, seu de Prudentia civili.*

21. Horace: Ars Poetica, 333. "Poets wish to profit or to please." Cardano has written *et . . . et for aut . . . aut*.

22. A favorite proverb of Cardano's: "A lion is known by his claws."

23. Paralipomena: "things omitted." This work Waters calls "the last fruit off an old tree." Morley: "a final heaping up, before he died, of all the chips that remained in his workshop." The collection included works on Algebra, Medicine, Natural History, Mechanics and Speculative Philosophy. They fill almost the entire 10th volume of the Opera Omnia, and in that edition also appeared in print for the first time.

24. Cardano apparently means his book *On the Best Way of Life: De Optimo Vitæ Genere,* mentioned at the beginning of the above discussion.

25. The group of his carefully written works.

26. This book: the *De Vita Propria Liber.*

27. The Ethics, probably. There is a break in the thought at this point and the Latin text offers considerable difficulty. Undoubtedly some sentences or phrases are missing.

28. The Politics, probably.

29. The work of Hippocrates: *Regimen in Acute Diseases.*

30. *Zedoary*: a medicinal substance obtained in the East Indies, having a fragrant smell, and a warm, bitter, aromatic taste. It is used in medicine as a stimulant.

31. *Lege Augustana*: What sort of gentleman's agreement this refers to I do not know.

CHAPTER FORTY-SIX

1. II Kings: 20: 6, "And I will add unto thy days fifteen years; and I will deliver thee and this city. . . ."

CHAPTER FORTY-SEVEN

1. The classical Latin word would be *genius.*

2. Alluding to the excellent treatment Josephus received as a prisoner, at the hands of Vespasian, Titus, and Domitian, to his experience with his fanatical fellow countrymen in the cave near Iotapata in Galilee, and to his shipwreck on the journey to Rome.

3. Cardano tells the story of the mysterious visitant who addressed him with the words *Te Sin casa* in Chapter 43. What the other incidents and "*Lamant*" refer to I have not been able to discover.

CHAPTER FORTY-EIGHT

(The numbers from 2 on correspond to numbers of names in chapter. Names sufficiently explained in the text, or previously in the notes, are omitted.)

1. *Corti*: See Chapter 14, note 6.

1. *Delfino*: See Chapter 30.

1. *Montana*: (Giambattista) famous Italian physician, 1498–1551.

1. *Camuzio* (Andrea): See Chapter 12, note 1.

1. *Crangius*: Not known.

2. *Aleman*: French physician of 16th Century.

5. *Ferrier*: French physician, 1513–1588.

6. *Hermes Trimaximus*: Egyptian author of the 2nd Century A.D.

7. *Mizaud*: French Astrologer, 1510–1578: *Catalogi sympathiæ et antipathiæ rerum aliquot memorabilium.*

8. *Amato*: *João Roderigo de Castello Branco:* Portuguese physician, 1510–1568.

9. *Andrea Baccio*: Italian physician of 15th Century.

11. *Antonio Maria de Conti*: Italian humanist, 1514–1555.

12. *Adrianus Turnebus*: a French classical scholar, 1512–1565, and specialist in Greek; Collège de France "Royal Reader."

13. *Jean Brodeau*: a French scholar, 1500–1568, and author of *"Dix Livres de Mélanges."*

14. *Buteo*: (*Jean Borrel*), French mathematician, 1492–1572.

15. *Charles de l'Ecluse*: a noted French botanist, 1524–1609. The complete title of his treatise is *Aromatum et Simplicium Aliquot Medicamentorum apud Indos Nascentium Historia.*

16. Uncertain.

17. *Bugatto*: an Italian annalist of the 16th Century.

18. *Sansovino*: (*Francesco Tatti*), Italian scholar, 1521–1586.

19. *Gesner*, the physician of Zurich, 1516–1565, author of *Bibliotheca Universalis.* Founder of modern study of Natural Science.

20. *Lycosthenes* (*Konrad Wolffhart*), German philological and theological author, 1518–1561. The title of his work is *Prodigiorum ac Ostentorum Chronicon.*

21. *Robert Constantine*: French physician and botanist, 1502–1603. He was a friend of Scaliger, and annotated one of the latter's commentaries on the Works of Theophrastus.

22. *Christoph Clavius*: 1537–1612, who took an active part in the Gregorian Calendar reform.

24. *Daniel Santberchius*: a Dutch mathematician.

25. Unknown.

26. Unknown.

27. *Francesco Alessandri*: Italian physician of the 16th Century.

28. *de Foix*: French mathematician, 1502–1594.

29. *Vicomercati*: Cardano's good friend (see Chapter 15). He was an Italian scholar of the period.

30. *Fuchs* (Leonhard): Professor of Medicine in University of Tübingen.

31. *Kaspar Peucer*: German physician and historian. He was Melanchthon's son-in-law.

33. *Pictorius*: George Pictor, an Alsatian physician, contemporary of Cardano, and author of many medical works similar to Cardano's.

35. *Gabriele Fallopio*: a noted Italian anatomist, 1523–1562.

36. *Rondelet*: French naturalist, 1507–1566.

37. *Frisius (Rainer)*: a physician and mathematician of the Low Countries. His Arithmetic was called *Methodus Arithmeticæ Practicæ.*

38. *Castiglione*: Girolamo, president of the Milanese Senate.

39. *Tragus=Buck*, a German botanist. His work was called *Neues Kreutterbuch.*

40. *Monteux*: an illustrious French physician of the 16th Century, and personal physician to Henry II of France.

41. *Peletier*: French poet and mathematician, 1517–1582.

42. *Jean Duchoul*: French naturalist.

43. *De Collado* (*Juan Luiz*): a Spanish physician of 16th Century,

and personal physician to Philip II. He wrote a Commentary on Galen's work.

44. *Plozio=de Plotis*, an Italian jurist of the period.

45. *Johann Schöner*: German astronomer and geographer, 1477–1547.

46. *Johann Cochläus* (Dobneck) was a German theologian and humanist. His *Historia* was called *De Actis et Scriptis Luthericis*.

47. *Scelerus* seems to be unknown, but *Juan de Seville* was a Spanish physician of the 15th Century.

49. *Stadius* was a Belgian mathematician. He wrote *Tabulæ Æquabilis et Apparentis Motus Corporum Cælestuim* and *Ephemerides Astrologicœ novæ*.

50. *João de Barros* was a Portuguese writer on historical subjects. The work referred to is *Quarta Decada da Asia*.

51. The great *Scaliger*, celebrated Italian scholar, 1494–1558.

52. *Charpentier*: French philosopher and physician, 16th Century.

53. *Giovanni Filippo Ingrassian*, a physician of Sicily, 16th Century.

54. Unknown.

55. Unknown.

56. *Lemmens*: philosopher and naturalist of the Low Countries. His book was called *De Occultis Naturæ Miraculis*.

57. Unknown.

58. *Suavius* (Jacques Gohary le Solitaire): French naturalist and poet.

59. *Gaurico*: Italian mathematician and astrologer (1476–1558) who wrote a *Tractatus Astrologicus*.

60. Unknown.

61. *Martinus Henricus*, author of *Quæstiones Medicas et consilium de diabeto.* Pavia, 1567.

62. *Philip Melanchthon*: the celebrated German reformer, 1497–1560.

63. Uncertain.

64. Uncertain.

65. *Bombelli* (Michael Raffaelle): an Italian mathematician, and specialist in algebra.

66. *Turtuglia*: See Chapter 44. *Niccolò Fontana*.

67. *Philandrier* (Guillaume): French scholar, 1505–1565.

68. *Pena*: French physician of 16th Century.

Lobel: French botanist.

69. *Solenander*: a German physician, who wrote *Tractatus de Causa Caloris Fontium Medicatorum Eorumque Temperatione*.

70. Uncertain.

71. *Duno*: an Italian physician, 1523–1601.

72. *Valentin Naibod*: mathematician of the 16th Century.

73. Unknown.

74. "A man of most profound, most favored and incomparable genius."

CHAPTER FORTY-NINE

1. This curious statement offers more difficulty in the matter of interpretation than translation.

2. Æneid: Book II, 774: "I quailed, my hair rose and I gasped for fear."

CHAPTER FIFTY

1. Ælius Donatus, the grammarian and commentator on *Vergil*.

2. *Vergil:* Eclogue V, 9: "Who cares if the fellow strive to rival Phoebus himself in song?"

3. Juvenal XIII, 20–23: "Philosophy is the conqueror of fortune, but they too are to be deemed happy who have learned under the schooling of life to endure its ills without fretting against the yoke."

4. Oh, hapless Virtue, thou good of words alone,
Why bendest thou thy course at Fortune's high demands?

5. Horace: Ars Poetica, 344: (A poet succeeds) "by pleasing and instructing his reader at the same time."

6. Ibid: 323, 324: "To the Greeks the Muse gave rounded phrase."

7. The Rhetoric.

8. "freedom from pain," i.e., *indolentia*.

CHAPTER FIFTY-ONE

1. *Sero sapiunt Phryges*: A proverb which originated with the Trojans who finally, after ten years, began to wish Helen and everything which had been stolen with her would return to the Greeks.

2. Horace: Odes III, 29, 43–44.

> I rest well pleased with former days;
> Let God from Heaven to-morrow give
> Black clouds or sunny rays.
>
> *Sir John Beaumont.*

CHAPTER FIFTY-TWO

1. Cardano was indeed thirty-nine years old before he overcame his earlier handicaps and began to live with some degree of comfort and prosperity.

2. Faustus, son of the Dictator and his fourth wife, Cæcilia Metella. He married Pompey's daughter and sided with his father-in-law in the Civil War. He was murdered in Cæsar's camp, whither he had been taken as prisoner, by the soldiers in a tumult.

3. Terentia is said to have lived to be 103 years old.

4. The lost works of the Philosopher, no doubt. Of the extant work of Theophrastus many editions appeared during Cardano's lifetime. Cardano's enemy, J. C. Scaliger, was the most accurate and brilliant scholar who has contributed to the elucidation of Theophrastus.

5. Jubellius Taurea, a Campanian of high rank and distinguished bravery in the second Punic war. He fought with Claudius Asellus in single combat in 215 B.C., and put an end to his own life on the capture of Capua by the Romans in 211 B.C.

CHAPTER FIFTY-THREE

1. Niccolò Fontana: Tartaglia, see Chapter 44.

2. Persius, Sat. I, 27. "Your knowledge is for naught unless another knows you know."

3. This last sentence offers considerable difficulty of interpretation, owing to a slight textual difficulty. The variant readings are "servatur" and "servat ut," neither of which is exactly the expression for the place.

CHAPTER FIFTY-FOUR

1. Chapter 40, *"Felicitas in curando."*

2. To advertise himself as a physician again ready to engage in the active practice of medicine.

BIBLIOGRAPHY

1. Bedingfield, T., *Cardanus' Comforte* (De Consolatione) translated into Englishe, and published by commaundement of the right honorable the Earle of Oxenford. T. Marche: London, 1573.

Same: Newly corrected and augmented. T. Marche, London, 1576.

Same: Another translation, called: *Cardan, his three books of Consolation Englished.* London, 1683.

2. Burr, Anna Robeson, *The Autobiography.* Houghton Mifflin Company, 1909.

3. Cantor, Moritz, *Hieronymus Cardanus: Ein wissenschaftliches Lebensbild aus dem 16 Jahrhunderts*: Neue Heidelberger Jahrbücher XIII, 1901–5, Vols. 11–13. Mitteilung an den Historiker Kongress in Rom im 1903.

4. Cardano, Girolamo, *Opera Omnia Hieronymi Cardani, Mediolanensis*: 10 vols. Spon: Lyons, 1663.

5. Crossley, J., *The Life and Times of Cardan.* London, 1836.

6. De Thou, M., *Histoire Universelle,* Vol. 7, p. 361.

7. Dunn, Waldo H., *English Biography,* Channels of English Literature. J. M. Dent & Sons, London, 1916.

8. Durey, Louis, *La Médecine Occulte de Paracelse et de quelques autres Médecins Hermétistes*: *Arnault de Villeneuve, J. Cardan, Cornelius Agrippa.* Vigot Freres: Paris, 1900.

9. *Firmiani, Girolamo Cardano, la vite e l'opere*: Naples, 1904.

10. Garnett, Richard, *Jerome Cardan.* Encyc. Brit. 11th Edit.

11. Gegenbauer, Leopold, *Sur . . . le cas irreductible de la formule de Cardane*: Soc. roy. d. sci. de Liège; mém. ser. 3. v. 2. Bruxelles, 1900.

12. Hallam's *Literature of Europe,* Vol. I, p. 395.

13. Hefele, Hermann, *Des Girolamo Cardano von Mailand* (Bürgers von Bologna) Eigene Lebensbeschreibung: Übertragen und eingeleitet. Eugen Diederichs: Jena, 1914.

14. Johnson, James C., *Biography: The Literature of Personality.* The Century Company, New York, 1927.

15. Kortholt, Christian, *De Tribus Impostoribus Magnis . . . et appendix, qua H. Cardani opiniones examinantur.* J. Reumannus. Kiel, 1680. Hamburg, 1701.

16. Lilly, Wm., *The Astrologer's Guide,* 1675. Choicest Aphorisms of the Seven Segments of Jerome Cardano of Milan. G. Redway, London, 1886.

Same, Gnostic Press, San Diego, Calif. 1918.

17. Mantovani, V., *Vita di Girolamo Cardano, Milanese, filosofo, medico e letterato celebratissimo,* scritta per lui medissimo in idioma latino e recanta nel italiano dal Sig. Dottore V. Mantovani. G. B. Sonzogno. Milan, 1821.

18. Morley, Henry, *The Life of Girolamo Cardano of Milan, Physician.* 2 vols. Chapman & Hall, London, 1854.

19. Naudè, Gabriel, *Hieronymi Cardani De Propria Vita Liber et G. Naudaei de Cardano Judicium.* Villery: Paris, 1643, and Ravestein: Amsterdam, 1654.

20. Parker, Samuel, *Tractatus de Deo.* 1678. (Includes Cardano among the Atheists.)

21. Rixner & Siber, *Leben und Lehrmeinungen berühmter Physiker am Ende des XVI und am Anfang des XVII Jahrhunderts.* Vol. II. Sulzbach, 1820.

22. Sardou, Victorien, *Nouvelle biographie génerale.* Vol. 8.

23. Scaliger, Julius Cæsar, *Exercitationum Esotericarum Libri xv. de Subtilitate ad H. Cardanum.* Paris, 1554.

24. Scaliger, J. C., *Epistolae aliquot unc primum vulgatae, etc.* R. Colomerii, Toulouse, 1620.

(N.B.—No. 23 is an attack upon Cardano; No. 24 an encomium of him.)

25. Symonds, J. A., *Renaissance in Italy.* John Murray, London. Vol. 2. 1923.

26. Tiraboschi, *Letteratura Italiana*, Tom. VII, Pt. 2, p. 623ff. Venice, 1824 (?).

27. Waters, W. G., *Jerome Cardan*, A Biographical Study. Lawrence & Bullen, Ltd. London, 1898.

28. Waters, W. G., *Pioneers in Humanism*, Anglo-Italian Review, 1919. Jan.–Apr.

29. Articles by Argellati, Biblio. Mediolan.; Bayle; Butrini; Grasset, M. Joseph; Italian Literary Historians: Ancona, Bartoli, Settembrini, Cantu; Lelut (Le Démon de Socrate); Leroux; Lombroso; Ribot (Psychologie des Sentiments), Retrospective Review; Introductory notices to the De Vita Propria Liber.

30. Cajori, F., *History of Mathematics.* University of California Press, 1919.

31. Guillaume Libri, *Histoire des Sciences Mathematiques en Italie depuis la Renaissance.* 1840. (Vol. 3, pp. 167–177.)

32. Cossali, *Storia dell'algebra.* Tom. II, p. 337 et seq.

ABOUT THE TYPE

The text of this book has been set in Trump Mediaeval. Designed by Georg Trump for the Weber foundry in the late 1950s, this typeface is a modern rethinking of the Garalde Oldstyle types (often associated with Claude Garamond) that have long been popular with printers and book designers.

TITLES IN SERIES

J. R. ACKERLEY Hindoo Holiday
J. R. ACKERLEY My Dog Tulip
J. R. ACKERLEY My Father and Myself
J. R. ACKERLEY We Think the World of You
DANTE ALIGHIERI The New Life
W. H. AUDEN (EDITOR) The Living Thoughts of Kierkegaard
HONORÉ DE BALZAC The Unknown Masterpiece *and* Gambara
MAX BEERBOHM Seven Men
ALEXANDER BERKMAN Prison Memoirs of an Anarchist
CAROLINE BLACKWOOD Corrigan
CAROLINE BLACKWOOD Great Granny Webster
MALCOLM BRALY On the Yard
ROBERT BURTON The Anatomy of Melancholy
CAMARA LAYE The Radiance of the King
GIROLAMO CARDANO The Book of My Life
J. L. CARR A Month in the Country
JOYCE CARY Herself Surprised (First Trilogy, Vol. 1)
JOYCE CARY To Be a Pilgrim (First Trilogy, Vol. 2)
JOYCE CARY The Horse's Mouth (First Trilogy, Vol. 3)
NIRAD C. CHAUDHURI The Autobiography of an Unknown Indian
ANTON CHEKHOV Peasants and Other Stories
COLETTE The Pure and the Impure
IVY COMPTON-BURNETT A House and Its Head
IVY COMPTON-BURNETT Manservant and Maidservant
JULIO CORTÁZAR The Winners
ASTOLPHE DE CUSTINE Letters from Russia
LORENZO DA PONTE Memoirs
ELIZABETH DAVID A Book of Mediterranean Food
ELIZABETH DAVID Summer Cooking
MARIA DERMOÛT The Ten Thousand Things
ARTHUR CONAN DOYLE Exploits and Adventures of Brigadier Gerard
CHARLES DUFF A Handbook on Hanging
J. G. FARRELL Troubles
M. I. FINLEY The World of Odysseus
MAVIS GALLANT Paris Stories
EDWARD GOREY (EDITOR) The Haunted Looking Glass
PETER HANDKE A Sorrow Beyond Dreams
ELIZABETH HARDWICK Seduction and Betrayal
ELIZABETH HARDWICK Sleepless Nights
L. P. HARTLEY Eustace and Hilda: A Trilogy
L. P. HARTLEY The Go-Between
JAMES HOGG The Private Memoirs and Confessions of a Justified Sinner
RICHARD HUGHES A High Wind in Jamaica
RICHARD HUGHES The Fox in the Attic (The Human Predicament, Vol. 1)
RICHARD HUGHES The Wooden Shepherdess (The Human Predicament, Vol. 2)

HENRY JAMES The Other House
HENRY JAMES The Outcry
RANDALL JARRELL (EDITOR) Randall Jarrell's Book of Stories
ERNST JÜNGER The Glass Bees
GEORG CHRISTOPH LICHTENBERG The Waste Books
JAMES MCCOURT Mawrdew Czgowchwz
HENRI MICHAUX Miserable Miracle
NANCY MITFORD Madame de Pompadour
ALBERTO MORAVIA Boredom
ALBERTO MORAVIA Contempt
ÁLVARO MUTIS The Adventures and Misadventures of Maqroll
L. H. MYERS The Root and the Flower
DARCY O'BRIEN A Way of Life, Like Any Other
IONA AND PETER OPIE The Lore and Language of Schoolchildren
BORIS PASTERNAK, MARINA TSVETAYEVA, AND RAINER MARIA RILKE Letters: Summer 1926
CESARE PAVESE The Moon and the Bonfires
CESARE PAVESE The Selected Works of Cesare Pavese
ANDREI PLATONOV The Fierce and Beautiful World
J. F. POWERS Morte d'Urban
J. F. POWERS The Stories of J. F. Powers
J. F. POWERS Wheat That Springeth Green
JEAN RENOIR Renoir, My Father
FR. ROLFE Hadrian the Seventh
WILLIAM ROUGHEAD Classic Crimes
DANIEL PAUL SCHREBER Memoirs of My Nervous Illness
JAMES SCHUYLER Alfred and Guinevere
LEONARDO SCIASCIA To Each His Own
LEONARDO SCIASCIA The Wine-Dark Sea
SHCHEDRIN The Golovlyov Family
TESS SLESINGER The Unpossessed: A Novel of the Thirties
CHRISTINA STEAD Letty Fox: Her Luck
STENDHAL The Life of Henry Brulard
ITALO SVEVO As a Man Grows Older
A. J. A. SYMONS The Quest for Corvo
EDWARD JOHN TRELAWNY Records of Shelley, Byron, and the Author
LIONEL TRILLING The Middle of the Journey
IVAN TURGENEV Virgin Soil
ROBERT WALSER Jakob von Gunten
ROBERT WALSER Selected Stories
SYLVIA TOWNSEND WARNER Lolly Willowes
SYLVIA TOWNSEND WARNER Mr. Fortune's Maggot *and* The Salutation
GLENWAY WESCOTT The Pilgrim Hawk
PATRICK WHITE Riders in the Chariot